SpringerBriefs in Materials

Series Editors

Sujata K. Bhatia, University of Delaware, Newark, USA

Alain Diebold, Schenectady, USA

Juejun Hu, Department of Materials Science and Engineering, Massachusetts Institute of Technology, Cambridge, USA

Kannan M. Krishnan, University of Washington, Seattle, USA

Dario Narducci, Department of Materials Science, University of Milano Bicocca, Milano, Italy

Suprakas Sinha Ray, Centre for Nanostructures Materials, Council for Scientific and Industrial Research, Brummeria, Pretoria, South Africa

Gerhard Wilde, Altenberge, Germany

The SpringerBriefs Series in Materials presents highly relevant, concise monographs on a wide range of topics covering fundamental advances and new applications in the field. Areas of interest include topical information on innovative, structural and functional materials and composites as well as fundamental principles, physical properties, materials theory and design.

Indexed in Scopus (2022).

SpringerBriefs present succinct summaries of cutting-edge research and practical applications across a wide spectrum of fields. Featuring compact volumes of 50 to 125 pages, the series covers a range of content from professional to academic. Typical topics might include

- A timely report of state-of-the art analytical techniques
- A bridge between new research results, as published in journal articles, and a contextual literature review
- A snapshot of a hot or emerging topic
- An in-depth case study or clinical example
- A presentation of core concepts that students must understand in order to make independent contributions

Briefs are characterized by fast, global electronic dissemination, standard publishing contracts, standardized manuscript preparation and formatting guidelines, and expedited production schedules.

Pierre Ramos · Luis Sánchez · Juan Rodríguez

Enhancing Photocatalytic Water Decontamination

Exploring the Efficacy of ZnO Nanorods in Advanced Treatment Processes

Pierre Ramos
Center for the Development of Advanced
Materials and Nanotechnology
Universidad Nacional de Ingeniería
Lima, Peru

Luis Sánchez
Center for the Development of Advanced
Materials and Nanotechnology
Universidad Nacional de Ingeniería
Lima, Peru

Juan Rodríguez
Center for the Development of Advanced
Materials and Nanotechnology
Universidad Nacional de Ingeniería
Lima, Peru

ISSN 2192-1091   ISSN 2192-1105 (electronic)
SpringerBriefs in Materials
ISBN 978-3-031-82505-7   ISBN 978-3-031-82506-4 (eBook)
https://doi.org/10.1007/978-3-031-82506-4

© The Editor(s) (if applicable) and The Author(s), under exclusive license to Springer Nature Switzerland AG 2025

This work is subject to copyright. All rights are solely and exclusively licensed by the Publisher, whether the whole or part of the material is concerned, specifically the rights of translation, reprinting, reuse of illustrations, recitation, broadcasting, reproduction on microfilms or in any other physical way, and transmission or information storage and retrieval, electronic adaptation, computer software, or by similar or dissimilar methodology now known or hereafter developed.

The use of general descriptive names, registered names, trademarks, service marks, etc. in this publication does not imply, even in the absence of a specific statement, that such names are exempt from the relevant protective laws and regulations and therefore free for general use.

The publisher, the authors and the editors are safe to assume that the advice and information in this book are believed to be true and accurate at the date of publication. Neither the publisher nor the authors or the editors give a warranty, expressed or implied, with respect to the material contained herein or for any errors or omissions that may have been made. The publisher remains neutral with regard to jurisdictional claims in published maps and institutional affiliations.

This Springer imprint is published by the registered company Springer Nature Switzerland AG
The registered company address is: Gewerbestrasse 11, 6330 Cham, Switzerland

If disposing of this product, please recycle the paper.

Contents

1 **Introduction** 1
   References 2
2 **Synthesis Methods for ZnO Nanorods** 5
  2.1 ZnO NRs Synthesized by Sol–Gel Method 5
  2.2 Gas-Phase Growth One-Dimensional ZnO Nanostructures 7
   References 11
3 **Mechanisms of ZnO Photocatalysis** 15
4 **Methods to Enhance ZnO Nanorods Photocatalytic Activity** 17
  4.1 ZnO Nanorods Doped with Metals or Nonmetals 17
     4.1.1 ZnO Nanorods Doped with Transition Metals 17
     4.1.2 ZnO Nanorods Doped with Rare Earth Metals 20
     4.1.3 Deposition of Noble Metals Onto ZnO Nanorods 20
     4.1.4 ZnO Nanorods Doped with Nonmetals 22
  4.2 Heterojunctions 24
  4.3 ZnO Nanorods-Based Carbonaceous Materials 29
     4.3.1 Application of ZnO Nanorods-Based Carbonaceous Photocatalyst for Dye Degradation 33
  4.4 ZnO Ternary Composites 40
   References 43
5 **Summary and Outlook for Future Directions** 51

Chapter 1
Introduction

Zinc oxide (ZnO) is a significant n-type semiconductor material known for its exceptional performance across various fields, including sensors, energy, nanomedicine, optoelectronics, and environmental applications [1]. ZnO possesses remarkable properties, such as a wide optical band gap of 3.37 eV, a high exciton binding energy of 60 meV, and an optical gain of 300 cm^{-1} at room temperature [2, 3]. Additionally, ZnO typically crystallizes in the hexagonal wurtzite structure and can form an array of nanostructures, including nanowires, nanorings, nanofibers, nanotubes, nanospheres, and nanorods [4–8]. These nanostructures demonstrate unique optical, electronic, physical, and chemical properties due to their large surface area, compatibility, abundance, non-toxicity, and ease of synthesis [9–12]. Among the various ZnO nanostructures, ZnO nanorods (NRs) have gained particular interest recently, as they provide excellent crystalline quality, high electron mobility, a large surface-to-volume ratio, high porosity, potential for quantum confinement [13].

In recent decades, various physical and chemical methods have been developed for synthesizing ZnO nanorods, including molecular beam epitaxy [13], sputtering [13, 14], chemical vapor deposition [15], electrodeposition [16], sol–gel processes [17, 18], and hydrothermal techniques [19, 20]. Among these, aqueous solution-phase growth methods, particularly those utilizing a hydrothermal approach, have gained significant attention due to their cost-effectiveness, simplicity of experimental setup, and potential for large-scale production [21–23]. The most common strategy for achieving well-controlled morphology and growth direction of ZnO NRs through hydrothermal methods involves using seeded substrates—substrates with a ZnO seed layer deposited by various techniques, including, spin-coating [24], dip-coating [25], electrospinning [26], spray pyrolysis [27], and sputtering [28]. These coated substrates reduce the thermodynamic barrier and provide nucleation centers promoting ZnO nanorod growth.

In the field of photocatalysis, ZnO has gained significant attention for its effectiveness in treating wastewater and pollutants [29, 30]. However, due to its wide optical band gap, ZnO primarily absorbs UV light, which accounts for only about

4% of the solar spectrum, leaving it unable to utilize 43% of the solar spectrum corresponding to visible light [31]. This limitation reduces ZnO's effectiveness as a photocatalyst. Consequently, enhancing the photocatalytic activity of ZnO nanorods by extending their response to visible light and reducing electron–hole recombination has become a key research focus. Various strategies have been developed to address these challenges, including doping ZnO nanorods with metal or nonmetal elements [32–35], coupling them with carbon-based materials [36–38], the formation of heterostructures with other semiconductors [39–42], and recently the synthesis of ternary composites based in ZnO nanorods [43–45].

This book summarizes current approaches for enhancing the performance of one-dimensional ZnO nanorods in pollutant photodegradation in water. It highlights various modification strategies that have successfully improved the photodegradation of organic dyes and pharmaceutical products. The focus is on understanding the fundamental mechanisms underlying the enhancement of photocatalysis through these surface modification techniques.

References

1. G.C. Yi, C. Wang, and W.I. Park, ZnO nanorods: Synthesis, characterization and applications. Semicond. Sci. Technol. **20**(4). IOP Publishing, S22 (2005). https://doi.org/10.1088/0268-1242/20/4/003
2. A. Khayatian, M.A. Kashi, R. Azimirad, S. Safa, S.F.A. Akhtarian, Effect of annealing process in tuning of defects in ZnO nanorods and their application in UV photodetectors. Optik **127**(11), 4675–4681 (2016). https://doi.org/10.1016/j.ijleo.2016.01.177
3. J. Lee, J. Chung, S. Lim, Improvement of optical properties of post-annealed ZnO nanorods. Physica E **42**(8), 2143–2146 (2010). https://doi.org/10.1016/j.physe.2010.04.013
4. B. Shouli et al., Different morphologies of ZnO nanorods and their sensing property. Sens. Actuators, B Chem. **146**(1), 129–137 (2010). https://doi.org/10.1016/j.snb.2010.02.011
5. C. Li et al., Hexagonal ZnO nanorings: Synthesis, formation mechanism and trimethylamine sensing properties. RSC Adv. **5**(98), 80561–80567 (2015). https://doi.org/10.1039/c5ra14793j
6. V. Gerbreders et al., Hydrothermal synthesis of ZnO nanostructures with controllable morphology change. Cryst. Eng. Comm. **22**(8), 1346–1358 (2020). https://doi.org/10.1039/c9ce01556f
7. A. di Mauro, M. Zimbone, M.E. Fragalà, G. Impellizzeri, Synthesis of ZnO nanofibers by the electrospinning process. Mater. Sci. Semicond. Process. **42**, 98–101 (2016). https://doi.org/10.1016/j.mssp.2015.08.003
8. P. Samadipakchin, H.R. Mortaheb, A. Zolfaghari, ZnO nanotubes: Preparation and photocatalytic performance evaluation. J. Photochem. Photobiol., A **337**, 91–99 (2017). https://doi.org/10.1016/j.jphotochem.2017.01.018
9. M. Sheikh et al., Application of ZnO nanostructures in ceramic and polymeric membranes for water and wastewater technologies: A review, Chem. Eng. J. **391**. Elsevier B.V., p. 123475 (2020). https://doi.org/10.1016/j.cej.2019.123475
10. R. Ahmad, S.M. Majhi, X. Zhang, T.M. Swager, and K.N. Salama, Recent progress and perspectives of gas sensors based on vertically oriented ZnO nanomaterials, Adv. Colloid Interface Sci. **270**. Elsevier B.V., 1–27 (2019). https://doi.org/10.1016/j.cis.2019.05.006
11. C.R. Chandraiahgari et al., Synthesis and characterization of ZnO nanorods with a narrow size distribution. RSC Adv. **5**(62), 49861–49870 (2015). https://doi.org/10.1039/c5ra02631h

12. P.G. Ramos et al., Enhanced photoelectrochemical performance and photocatalytic activity of ZnO/TiO$_2$ nanostructures fabricated by an electrostatically modified electrospinning. Appl. Surf. Sci. **426**, 844–851 (2017). https://doi.org/10.1016/j.apsusc.2017.07.218
13. R. Nandi, S.S. Major, The mechanism of growth of ZnO nanorods by reactive sputtering. Appl. Surf. Sci. **399**, 305–312 (2017). https://doi.org/10.1016/j.apsusc.2016.12.097
14. C. Baratto, Growth and properties of ZnO nanorods by RF-sputtering for detection of toxic gases. RSC Adv. **8**(56), 32038–32043 (2018). https://doi.org/10.1039/c8ra05357j
15. Q. Zhang, C. Li, TiO$_2$ coated zno nanorods by mist chemical vapor deposition for application as photoanodes for dye-sensitized solar cells. Nanomaterials **9**(9), 1339 (2019). https://doi.org/10.3390/nano9091339
16. S.N. Sarangi, Controllable growth of ZnO nanorods via electrodeposition technique: Towards UV photo-detection. J. Phys. D Appl. Phys. **49**(35), 355103 (2016). https://doi.org/10.1088/0022-3727/49/35/355103
17. F. Aslan, A. Tumbul, A. Göktaş, R. Budakoğlu, İH. Mutlu, Growth of ZnO nanorod arrays by one-step sol–gel process. J. Sol-Gel Sci. Technol. **80**(2), 389–395 (2016). https://doi.org/10.1007/s10971-016-4131-z
18. K.L. Foo, U. Hashim, K. Muhammad, C.H. Voon, Sol–gel synthesized zinc oxide nanorods and their structural and optical investigation for optoelectronic application. Nanoscale Res. Lett. **9**(1), 1–10 (2014). https://doi.org/10.1186/1556-276X-9-429
19. S.J. Young, C.L. Chiou, Synthesis and optoelectronic properties of Ga-doped ZnO nanorods by hydrothermal method. Microsyst. Technol. **24**(1), 103–107 (2018). https://doi.org/10.1007/s00542-016-3183-x
20. M. Poornajar, P. Marashi, D. Haghshenas Fatmehsari, and M. Kolahdouz Esfahani, Synthesis of ZnO nanorods via chemical bath deposition method: The effects of physicochemical factors, Ceram. Int. **42**(1), 173–184 (2016). https://doi.org/10.1016/j.ceramint.2015.08.073
21. N.A. Alshehri, A.R. Lewis, C. Pleydell-Pearce, T.G.G. Maffeis, Investigation of the growth parameters of hydrothermal ZnO nanowires for scale up applications. J. Saudi Chem. Soc. **22**(5), 538–545 (2018). https://doi.org/10.1016/j.jscs.2017.09.004
22. W. Wang et al., Fabrication and optical property of ZnO nanorod array by hydrothermal method. Ferroelectrics **549**(1), 204–211 (2019). https://doi.org/10.1080/00150193.2019.1592561
23. C.H. Huang, Y.L. Chu, L.W. Ji, I.T. Tang, T.T. Chu, and B.J. Chiou, Fabrication and characterization of homostructured photodiodes with Li-doped ZnO nanorods, Microsyst. Technol. pp 1–7 (2020). https://doi.org/10.1007/s00542-020-04854-1
24. M.Z. Toe et al., Effect of ZnO seed layer on the growth of ZnO nanorods on silicon substrate, Mater. Today: Proc. **17**, 553–559 (2019). https://doi.org/10.1016/j.matpr.2019.06.334
25. B. Nikola et al., Highly textured seed layers for the growth of vertically oriented ZnO nanorods. Crystals **9**(11), 566 (2019). https://doi.org/10.3390/cryst9110566
26. X. Dong, P. Yang, R. Shi, Fabrication of ZnO nanorod arrays via electrospinning assisted hydrothermal method. Mater. Lett. **135**, 96–98 (2014). https://doi.org/10.1016/j.matlet.2014.07.102
27. E. Karaköse, H. Çolak, Structural and optical properties of ZnO nanorods prepared by spray pyrolysis method. Energy **140**, 92–97 (2017). https://doi.org/10.1016/j.energy.2017.08.109
28. B.S. Sannakashappanavar, N.A. Pattanashetti, C.R. Byrareddy, and A.B. Yadav, Study of annealing effect on the growth of ZnO nanorods on ZnO seed layers, AIP Conference Proceedings, **1943**(1), 020077 (2018). https://doi.org/10.1063/1.5029653
29. K.M. Lee, C.W. Lai, K.S. Ngai, and J.C. Juan, Recent developments of zinc oxide based photocatalyst in water treatment technology: A review, Water Res. **88**. Elsevier Ltd, pp 428–448 (2016). https://doi.org/10.1016/j.watres.2015.09.045
30. M.A. Mohd Adnan, N.M. Julkapli, and S.B. Abd Hamid, Review on ZnO hybrid photocatalyst: Impact on photocatalytic activities of water pollutant degradation, Rev. Inorg. Chem. **36**(2). Walter de Gruyter GmbH, pp. 77–104 (2016). https://doi.org/10.1515/revic-2015-0015
31. Y. Zhang, R. Mandal, D.C. Ratchford, R. Anthony, J. Yeom, Si nanocrystals/ZnO nanowires hybrid structures as immobilized photocatalysts for photodegradation. Nanomaterials **10**(3), 491 (2020). https://doi.org/10.3390/nano10030491

32. R. Rooydell, S. Brahma, R.-C. Wang, M.R. Modaberi, F. Ebrahimzadeh, C.-P. Liu, Cu doped ZnO nanorods with controllable Cu content by using single metal organic precursors and their photocatalytic and luminescence properties. J. Alloy. Compd. **691**, 936–945 (2017). https://doi.org/10.1016/j.jallcom.2016.08.324
33. N.R. Khalid et al., Enhanced photocatalytic activity of Al and Fe co-doped ZnO nanorods for methylene blue degradation. Ceram. Int. **45**(17), 21430–21435 (2019). https://doi.org/10.1016/j.ceramint.2019.07.132
34. M. Koohgard, A.M. Sarvestani, M. Hosseini-Sarvari, Photocatalytic synthesis of unsymmetrical thiourea derivatives via visible-light irradiation using nitrogen-doped ZnO nanorods. New J. Chem. **44**(34), 14505–14512 (2020). https://doi.org/10.1039/d0nj02197k
35. V. Kumari, A. Mittal, J. Jindal, S. Yadav, and N. Kumar, S-, N- and C-doped ZnO as semiconductor photocatalysts: A review, Front. Mater. Sci. **13**(1). Higher Education Press (2019). https://doi.org/10.1007/s11706-019-0453-4
36. X. Ye, Z. Wang, Z. Li, W. Li, Q. Wang, ZnO nanorod array/reduced graphene oxide substrate with enhanced performance in photocatalytic degradation. Micro Nano Letters **14**(8), 868–871 (2019). https://doi.org/10.1049/mnl.2018.5446
37. R. Cai et al., 3D graphene/ZnO composite with enhanced photocatalytic activity. Mater. Des. **90**, 839–844 (2016). https://doi.org/10.1016/j.matdes.2015.11.020
38. H. Moussa et al., Growth of ZnO Nanorods on Graphitic Carbon Nitride gCN Sheets for the Preparation of Photocatalysts with High Visible-Light Activity. ChemCatChem **10**(21), 4987–4997 (2018). https://doi.org/10.1002/cctc.201801206
39. P. Senthil Kumar, M. Selvakumar, S. Ganesh Babu, S. Induja, and S. Karuthapandian, CuO/ZnO nanorods: An affordable efficient p-n heterojunction and morphology dependent photocatalytic activity against organic contaminants, J. Alloys Compounds. **701**, 562–573 (2017). https://doi.org/10.1016/j.jallcom.2017.01.126
40. H. Li et al., Enhanced photocatalytic activity and synthesis of ZnO nanorods/MoS_2 composites. Superlattices Microstruct. **117**, 336–341 (2018). https://doi.org/10.1016/j.spmi.2018.03.028
41. S.M. Lam, J.C. Sin, A.Z. Abdullah, A.R. Mohamed, Sunlight responsive WO_3/ZnO nanorods for photocatalytic degradation and mineralization of chlorinated phenoxyacetic acid herbicides in water. J. Colloid Interface Sci. **450**, 34–44 (2015). https://doi.org/10.1016/j.jcis.2015.02.075
42. M. Kwiatkowski, I. Bezverkhyy, M. Skompska, ZnO nanorods covered with a TiO_2 layer: Simple sol-gel preparation, and optical, photocatalytic and photoelectrochemical properties. J. Mat. Chem. A **3**(24), 12748–12760 (2015). https://doi.org/10.1039/c5ta01087j
43. L. Mohanty, D.S. Pattanayak, S.K. Dash, An efficient ternary photocatalyst Ag/ZnO/g-C_3N_4 for degradation of RhB and MG under solar radiation. J. Indian Chem. Soc. **98**(11), 100180 (2021). https://doi.org/10.1016/j.jics.2021.100180
44. K. Chaudhary et al., Binary WO_3-ZnO nanostructures supported rGO ternary nanocomposite for visible light driven photocatalytic degradation of methylene blue Synth. Met. **269**, 116526 https://doi.org/10.1016/j.synthmet.2020.116526
45. R. Gupta, N. K. Eswar, J. M. Modak, G. Madras, Effect of morphology of zinc oxide in ZnO-CdS-Ag ternary nanocomposite towards photocatalytic inactivation of E. coli under UV and visible light, Chem. Eng. J. **307**, 966–980 (2017). https://doi.org/10.1016/j.cej.2016.08.142

Chapter 2
Synthesis Methods for ZnO Nanorods

2.1 ZnO NRs Synthesized by Sol–Gel Method

This section reviews the sol–gel synthesis process for ZnO nanorods. The sol–gel method, also known as soft chemistry, is a wet chemical technique that enables the formation of solid materials from a colloidal suspension. This method is widely used for fabricating glasses, ceramics, thin films, and fibers, offering excellent control over the texture and surface properties of the resulting materials [1]. The sol–gel process typically involves six main steps: hydrolysis, condensation, polymerization, aging, drying, and thermal decomposition [2]. Figure 2.1 illustrates various sol–gel methods for synthesizing different material forms [3]. In 1990, Andres Verges et al. [4] published the first study on the formation of zinc oxide microcrystals using an aqueous solution growth method, which involves the hydrolysis of zinc nitrate and zinc chloride in the presence of hexamethylenetetramine. They discovered that the resulting structures, such as rods, needles, or spherulitic aggregates, depend on several factors, including reactant concentrations, pH, and temperature. Ram SDG et al. [5] proposed a mechanism for forming ZnO rods in the presence of hexamethylenetetramine:

$$(CH_2)_6N_4 + 6H_2O \rightarrow 4NH_3 + 6HCH \quad (2.1)$$

$$NH_3 + H_2O \rightarrow NH_4^+ + OH^- \quad (2.2)$$

$$Zn(NO_3)_2 \cdot 4H_2O + 2OH^- \rightarrow Zn(OH)_2 + 2NO_3^- + 4H_2O \quad (2.3)$$

$$Zn(OH)_2 \rightarrow Zn^{2+} + 2OH^- \quad (2.4)$$

$$Zn^{2+} + 2OH^- \rightarrow ZnO + H_2O \quad (2.5)$$

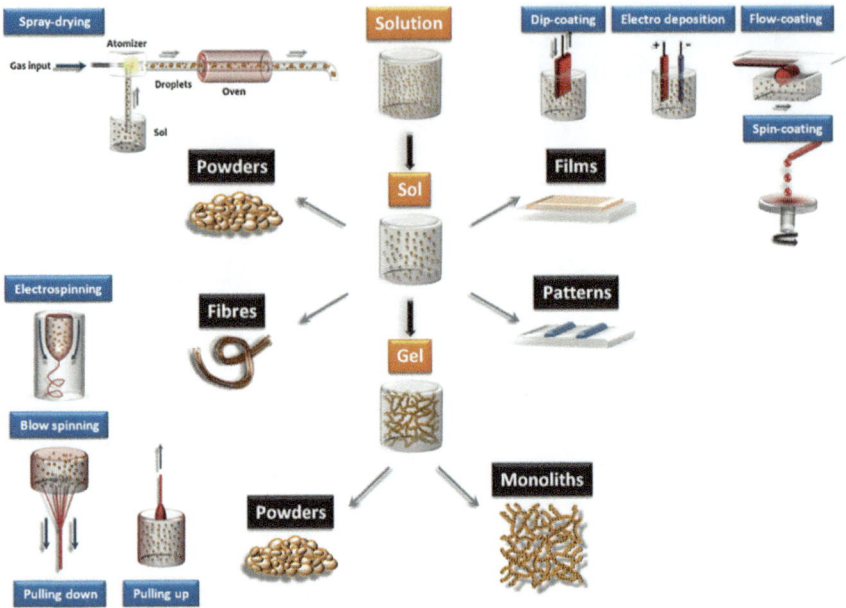

Fig. 2.1 Processing routes to materials using sol–gel methods [3]

More than a decade later, Vayssieres et al. [6] adopted the method developed by Andres Verges to grow microrods using an equimolar (0.1 M) aqueous solution of zinc nitrate and methenamine. This process was conducted at 95 °C for 1–10 h on various substrates, including conducting glass, silicon substrates, bare glass, and conducting plastic. Typically, ZnO nanorods are grown through a two-step process. First, a polycrystalline ZnO seed layer is coated onto different substrates using various sol–gel techniques. ZnO nanorods are then grown on the seed layer using the hydrothermal method. The resulting single-crystalline hexagonal rods, with well-defined crystallographic faces, grew perpendicularly in the [0001] direction and were organized into large, uniform arrays, regardless of the substrate type. The length of the microrods was determined by the synthesis duration, with an average growth rate of approximately 1 μm per hour at 95 °C. According to Vayssieres et al. [7], the simplest way to obtain ZnO nanorods is by reducing the overall concentration of reagents while maintaining a 1:1 ratio. Decreasing the precursor concentration by an order of magnitude results in a corresponding reduction in the rod diameter, due to critical monomer diffusion and limited growth.

Baruah and Dutta [8] investigated how pH variations affect the size and morphology of ZnO nanorods grown using hydrothermal processes at temperatures below 100 °C. They found that during the growth process, the pH of a solution containing zinc acetate dihydrate (ZAH) and hexamethylenetetramine (HMT), gradually shifted from 6.4 to 7.3 after five hours. The study found that ZnO nanorods grew faster in both width and length under acidic conditions, while at a pH of 8–12 (basic conditions), flower petal-like ZnO nanostructures formed. This suggests that starting with acidic conditions is essential for producing nanorods with well-defined hexagonal facets. However, because ZnO erodes in acidic environments, the final dimensions of the nanorods were determined by a balance between crystal growth and etching. Yin et al. [9] emphasized the crucial role of the seed layer in controlling the morphology and growth direction of ZnO nanorods. Additionally, the aqueous hydrothermal technique is considered an easy, cost-effective, energy-efficient, and convenient method for fabricating ZnO nanorods, particularly due to its capability for large-scale production at low growth temperatures. Heterogeneous nucleation supports the growth of ZnO nanorods, as the interfacial energy between the crystals and the solution is typically higher than between the crystals and the substrate [10].

The sol–gel fabrication of ZnO nanorods is influenced by several key factors [11, 12]:

(i) The preparation of the precursor solution, where the molar ratio of the zinc precursor to the solvent and additives affects the particle size and morphology.
(ii) The coating of the substrate with the prepared sol, where the chosen deposition technique significantly influences the alignment, density, and uniformity of the ZnO nanorods.
(iii) The heat treatment of the xerogel film, significantly impacts the size and shape of the ZnO.
(iv) The preparation of growth solution based on zinc precursor (usually zinc nitrate) and reducing agent (e.g. hexamethylenetetramine or sodium hydroxide).

2.2 Gas-Phase Growth One-Dimensional ZnO Nanostructures

Gas-phase methods for synthesizing one-dimensional ZnO nanostructures have been extensively employed for many years to generate high-quality materials efficiently [13–16]. In the gas-phase method, the vapor of a desired material is generated either by evaporation or chemical reduction and later, transported to a solid substrate surface by a gas carrier, where it condenses under specific conditions (temperature, pressure, atmosphere, substrates, etc.) to produce high-quality ZnO nanostructures, following the vapor–liquid-solid (VLS) or vapor–solid (VS) process. Gas phase methods, including chemical vapor deposition (CVD) and physical vapor deposition (PVD), are typically employed at high temperatures and low pressures and generally require the zinc precursor to be in the vapor state before the deposition. Both PVD and CVD convert a vapor of a given substance into a solid coating on the surface of a substrate. However, they differ in how the process occurs. PVD involves removing atoms, molecules, and ions from a solid source (target) and condensing them onto the substrate. CVD uses vaporized liquids or gases as sources that chemically react on or close to the surface of the substrate to form a solid film. While PVD processes always occur in a vacuum, CVD can arise in high vacuum to atmospheric pressure.

Rusli N. et al. [17] synthesized high-density ZnO nanorods on porous silicon (PS) substrates using straightforward thermal evaporation of Zn powder in the presence of O_2 gas within a single-zone horizontal tube furnace, as illustrated in Fig. 2.2 at growth temperatures ranging from 600 °C to 1000 °C. Metallic zinc powder and oxygen gas were utilized as the sources without any catalyst. A high density of ZnO nanorods was produced over an extensive region, and it has been attributed to the textured surface of PS, which offers suitable planes to facilitate the deposition of Zn or ZnO_x seeds, serving as nucleation sites for the ensuing growth of ZnO nanorods. The diameters and geometrical morphologies of ZnO nanorods produced via this single-step process are significantly affected by the structures of the ZnO_x seeds generated during the first growing phase. A combination of self-catalyzed VLS and VS processes characterizes the growth mechanism.

An alternative synthesis method to produce one-dimensional ZnO nanostructures is the application of catalyst-assisted growth, commonly known as the VLS process, which was initially developed in the 1960s to produce silicon microstructures [18]. This requires the deposition of metallic islands on the substrates before ZnO growth. An advantage of catalytic growth is the ability to control the position, density, and diameter of the one-dimensional nanostructures; nevertheless, the growth temperature needs to exceed the catalyst's melting point. Sangpour et. al. [19] synthesized one-dimensional ZnO nanostructures (nanowires, nanorods, and nanopillars) on a gold-coated silicon substrate by employing ZnO and carbon powders as reactants through a VLS process at temperatures in the range of 875 °C to 910 °C in an

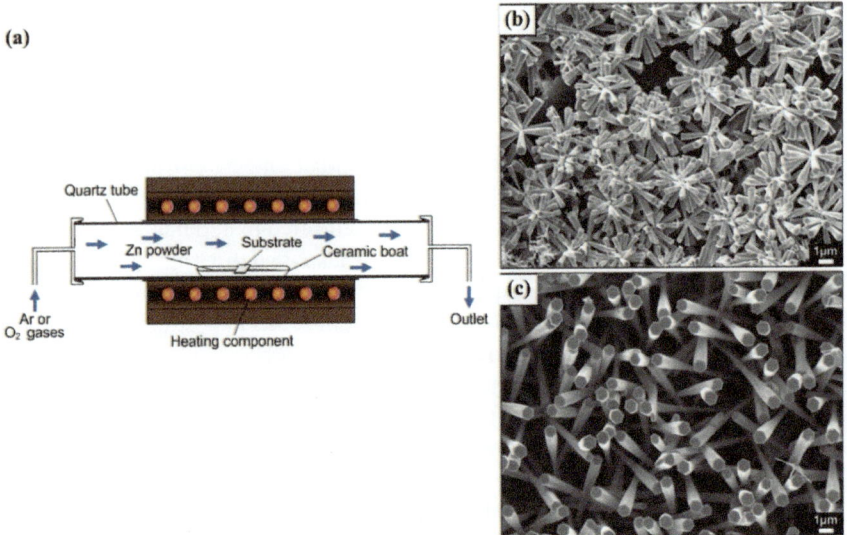

Fig. 2.2 a Schematic of a tube furnace system of nanorod growth by the thermal evaporation method. FESEM images of ZnO nanorods grown on porous silicon substrates at different growth temperatures of **b** 600 °C; and **c** 800 °C [17]

2.2 Gas-Phase Growth One-Dimensional ZnO Nanostructures

experimental setup similar to that shown in Fig. 2.2 a. The VLS growth mechanism of ZnO one-dimensional ZnO nanostructures which consists of three stages: metal thin film deposition, catalytic nanoparticles formation, ZnO nucleation, and nanowire epitaxial growth is shown in Fig. 2.3. After the Zn vapor was transported to the surface of the substrate, it initially formed liquid droplets that reacted with the existing gold catalyst on the surface to create an alloy that served as nucleation sites for the growth of one-dimensional ZnO nanostructures. The growth temperature, determined by the distance from the Zn source to the substrate, was identified as the crucial experimental parameter for the synthesis of various morphologies of ZnO nanostructures. ZnO nanowires were synthesized at the region of lowest temperature, the zone the furthest from the Zn source, while the nanopillars grew closer to the Zn source at the highest temperatures. ZnO nanorods with diameters of around 160 nm and lengths of up to 2 microns were synthesized at intermediate temperatures between nanowire and nanopillar nanostructures.

Zhang et. al. [20] examined the influence of catalysts on ZnO nanowire growth by contrasting the efficacy of Au, Pt, and Ag nanoparticles. An improved growth

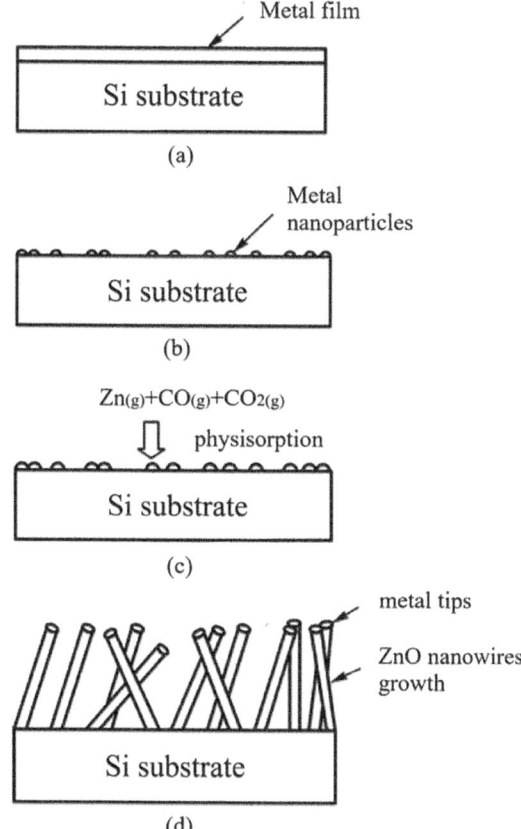

Fig. 2.3 Schematic illustration of VLS growth mechanism of ZnO nanowire: **a** Metal film deposition, **b** Au nanoparticles formation under heat treatment, **c** adsorption and nucleation, and **d** growth of nanowires. [19]

control was achieved by employing a substrate temperature of 800 °C with a ZnO/graphite powder source or 500 °C with a Zn powder source. At 800 °C, the VLS process is essential when the Au and Pt nanoparticles are in a liquid phase. In contrast, nanowires can be synthesized on solid nanoparticles through the vapor–solid (VS) mechanism, specifically from oxidized silver at 800 °C and nanoscale cracks of platinum at 500 °C.

Chemical vapor deposition (CVD) is an appealing method for the large-scale production of ZnO nanorods [21–23]. Kumar S. et al. [23] synthesized vertically aligned single crystalline ZnO nanorods arrays on a silicon substrate in a CVD reactor using Au as a catalyst. The CVD parameters, including substrate temperature, catalyst layer thickness/morphology, and reaction time, are essential in synthesizing nanostructures via the VLS process. The key factors identified for the growth and preventing seed poisoning, are a temperature of 800 °C for optimal growth, having an Au catalyst layer thickness of 5.0 nm.

Metal–Organic Chemical Vapor Deposition (MOCVD) is a technique derived from CVD that uses metal–organic compounds as precursors. These compounds contain metal–carbon bonds vaporized to deposit ZnO nanostructures [24–26]. ZnO growth by MOCVD typically employs dimethylzinc (DMZn) or diethylzinc (DEZn) as zinc precursors and nitrogen or argon as gas carriers [27–29]. Farid Falyouni et al. [30, 31] have reported on the growth of vertically well-aligned 1D ZnO nanostructure by MOCVD technique using dimethylzinc–triethylamine (DMZn–TEN) and N_2O as zinc and oxygen precursors, respectively, and helium as a gas carrier. They found that the choice of gas carrier and its flow velocity can significantly influence the formation of ZnO nanostructures. Achieving well-defined arrays of one-dimensional structures in the 640–780 °C temperature range was challenging when using DMZn-TEN and N_2O as precursors, with N_2 as the gas carrier. In contrast, ZnO nanorods and nanotubes were synthesized under identical growth circumstances when nitrogen was substituted with helium.

Chemical bath deposition (CBD) has garnered significant interest due to its simple experimental setup, low cost, non-hazardous nature, reproducibility, and potential for scalability. This method allows deposition on a wide range of substrates. In the context of CBD growth of ZnO nanorods, hexamethylenetetramine (HMTA) is particularly effective, ensuring high crystallinity and favorable morphological properties of ZnO nanorods compared to other reducing agents. Additionally, CBD is recognized as a high-performance, effective, and efficient method for fabricating various nanostructures. For instance, Thanh Son et al. [32] fabricated ZnO seed layers on a Si substrate using RF magnetron sputtering at room temperature, varying the layer thickness between 80 and 200 nm by controlling the sputtering time. They investigated the thickness-dependent surface morphology of the seed layer and its impact on the vertical growth of ZnO nanorod arrays. The CBD technique was employed to grow these ZnO nanorod arrays using HMTA as a reducing agent. The results revealed that dense, hexagonally-shaped ZnO nanorods were almost vertically aligned, with their average diameters closely matching the average grain sizes of the ZnO films. Furthermore, the average diameter and length of the ZnO nanorods generally increased with the thickness of the ZnO film, which is related to the variations in the average grain size and RMS roughness of the ZnO film depending on its thickness. These findings indicate that ZnO thin films function effectively as seed layers for the vertically aligned growth of ZnO nanorods during the subsequent solution synthesis step. The vertically grown ZnO nanorods demonstrated excellent crystalline and optical qualities.

On the other hand, Abdulrahman et al. [33] synthesized ZnO nanorods using a modified CBD method within a temperature range of 65–115 °C. They systematically investigated the impact of different growth temperatures on the surface morphology, top view, cross-section (including growth direction and length), structure, elemental chemical composition, and optical properties of the ZnO nanorods. This study successfully fabricated high-quality, vertically aligned ZnO nanorods on glass substrates using a cost-effective, low-temperature, and straightforward modified CBD method. Thus, the chemical bath deposition method is highly efficient and reliable due to its simplicity and the high quality of the resulting products, where the Hexamethylenetetramine plays a crucial role in initiating and accelerating the nanorod growth through multiple chemical reactions. Finally, the main chemical reactions that describe the formation of ZnO nanorods are consistent with those mentioned in the previous section (see Eqs. 2.1–2.5).

References

1. M. Parashar, V.K. Shukla, and R. Singh, Metal oxides nanoparticles via sol–gel method: A review on synthesis, characterization and applications, J. Mat. Sci. Mat. Elect. **31**(5), 3729–3749 (2020). https://doi.org/10.1007/S10854-020-02994-8
2. L. Znaidi, Sol–gel-deposited ZnO thin films: Sol–gel-deposited ZnO thin films: A review, **174**(1–3), 18–30 (2010). https://doi.org/10.1016/j.mseb.2010.07.001
3. C. Sanchez, P. Belleville, M. Popall, L. Nicole, Applications of advanced hybrid organic–inorganic nanomaterials: From laboratory to market. Chem. Soc. Rev. **40**(2), 696–753 (2011). https://doi.org/10.1039/C0CS00136H
4. M.A. Vergés, A. Mifsud, C.J. Serna, Formation of rod-like zinc oxide microcrystals in homogeneous solutions. J. Chem. Soc. Faraday Trans. **86**(6), 959–963 (1990). https://doi.org/10.1039/FT9908600959
5. S.D.G. Ram, G. Ravi, A. Athimoolam, T. Mahalingam, and M.A. Kulandainathan, Aqueous chemical growth of free standing vertical ZnO nanoprisms, nanorods and nanodiskettes with improved texture co-efficient and tunable size uniformity, App. Phys. A. **105**(4), 881–890 (2011). https://doi.org/10.1007/S00339-011-6518-6
6. L. Vayssieres, K. Keis, S.-E. Lindquist, A. Hagfeldt, Purpose-built anisotropic metal oxide material: 3D highly oriented microrod array of ZnO. J. Phys. Chem. B **105**(17), 3350–3352 (2001). https://doi.org/10.1021/jp010026s
7. L. Vayssieres, Growth of arrayed nanorods and nanowires of ZnO from aqueous solutions. Adv. Mater. **15**(5), 464–466 (2003). https://doi.org/10.1002/adma.200390108
8. S. Baruah, J. Dutta, PH-dependent growth of zinc oxide nanorods. J. Cryst. GrowthCryst. Growth **311**(8), 2549–2554 (2009). https://doi.org/10.1016/j.jcrysgro.2009.01.135
9. Y.T. Yin, W.X. Que, C.H. Kam, ZnO nanorods on ZnO seed layer derived by sol–gel process. J. Sol-Gel Sci. Technol. **53**(3), 605–612 (2009). https://doi.org/10.1007/s10971-009-2138-4

10. N.S. Ridhuan, K.A. Razak, Z. Lockman, A.A. Aziz, Structural and morphology of ZnO nanorods synthesized Using ZnO seeded growth hydrothermal method and its properties as UV sensing. PLoS ONE **7**(11), e50405 (2012). https://doi.org/10.1371/journal.pone.0050405
11. Z. Liu, J. Ya, and L.E, Effects of substrates and seed layers on solution growing ZnO nanorods, J. Solid State Electrochem. **14**(6), 957–963 (2009). https://doi.org/10.1007/s10008-009-0894-2
12. K. Harun, F. Hussain, A. Purwanto, B. Sahraoui, A. Zawadzka, A.A. Mohamad, Sol–gel synthesized ZnO for optoelectronics applications: A characterization review. Mater. Res. Express **4**(12), 122001 (2017). https://doi.org/10.1088/2053-1591/aa9e82
13. G. Jimenez-Cadena, E. Comini, M. Ferroni, A. Vomiero, G. Sberveglieri, Synthesis of different ZnO nanostructures by modified PVD process and potential use for dye-sensitized solar cells. Mater. Chem. Phys. **124**(1), 694–698 (2010). https://doi.org/10.1016/j.matchemphys.2010.07.035
14. N.K. Hassan, M.R. Hashim, M. Bououdina, One-dimensional ZnO nanostructure growth prepared by thermal evaporation on different substrates: Ultraviolet emission as a function of size and dimensionality. Ceram. Int. **39**(7), 7439–7444 (2013). https://doi.org/10.1016/j.ceramint.2013.02.088
15. Z.L. Wang, Zinc oxide nanostructures: growth, properties and applications. J. Phys. Condens. MatterCondens. Matter **16**(25), R829 (2004). https://doi.org/10.1088/0953-8984/16/25/R01
16. S.Y. Li, C.Y. Lee, T.Y. Tseng, Copper-catalyzed ZnO nanowires on silicon (1 0 0) grown by vapor–liquid–solid process. J. Cryst. GrowthCryst. Growth **247**(3–4), 357–362 (2003). https://doi.org/10.1016/S0022-0248(02)01918-8
17. N.I. Rusli, M. Tanikawa, M.R. Mahmood, K. Yasui, A.M. Hashim, Growth of high-density zinc oxide nanorods on porous silicon by thermal evaporation. Materials **5**(12), 2817–2832 (2012). https://doi.org/10.3390/ma5122817
18. R.S. Wagner, W.C. Ellis, Vapor-liquid-solid mechanism of single crystal growth. Appl. Phys. Lett. **4**(5), 89–90 (1964). https://doi.org/10.1063/1.1753975
19. L.H. Madkour, Environmental impact of nanotechnology and novel applications of nano materials and nano devices, Nanoelectron. Mater., Adv. Struct. Mater. **116**, 605–699, Springer (2019). https://doi.org/10.1007/978-3-030-21621-4_16
20. Z. Zhang, S.J. Wang, T. Yu, T. Wu, Controlling the growth mechanism of ZnO nanowires by selecting catalysts. J. Phys. Chem. C **111**(47), 17500–17505 (2007). https://doi.org/10.1021/jp075296a
21. C. Ge, H. Li, M. Li, C. Li, X. Wu, B. Yang, Synthesis of a ZnO nanorod/CVD graphene composite for simultaneous sensing of dihydroxybenzene isomers. Carbon **95**, 1–9 (2015). https://doi.org/10.1016/j.carbon.2015.08.006
22. M. Bai, M. Chen, X. Li, Q. Wang, One-step CVD growth of ZnO nanorod/SnO_2 film heterojunction for NO_2 gas sensor. Sens. Actuators, B Chem. **373**, 132738 (2022). https://doi.org/10.1016/j.snb.2022.132738
23. S. Kumar, P.D. Sahare, S. Kumar, Optimization of the CVD parameters for ZnO nanorods growth: Its photoluminescence and field emission properties. Mater. Res. Bull. **105**, 237–245 (2018). https://doi.org/10.1016/j.materresbull.2018.05.002
24. D.N. Montenegro, A. Souissi, C. Martínez-Tomás, V. Muñoz-Sanjosé, V. Sallet, Morphology transitions in ZnO nanorods grown by MOCVD. J. Cryst. GrowthCryst. Growth **359**(1), 122–128 (2012). https://doi.org/10.1016/j.jcrysgro.2012.08.038
25. M. Rosina, P. Ferret, P.H. Jouneau, I.C. Robin, F. Levy, G. Feuillet, M. Lafossas, Morphology and growth mechanism of aligned ZnO nanorods grown by catalyst-free MOCVD. Microelectron. J.. J. **40**(2), 242–245 (2009). https://doi.org/10.1016/j.mejo.2008.07.019
26. B. Wu, Y. Zhang, Z. Shi, X. Li, X. Cui, S. Zhuang, B. Zhang, G. Du, Different defect levels configurations between double layers of nanorods and film in ZnO grown on c-Al_2O_3 by MOCVD. J. Lumin.Lumin. **154**, 587–592 (2014). https://doi.org/10.1016/j.jlumin.2014.06.004
27. D.C. Kim, B.H. Kong, H.K. Cho, Morphology control of 1D ZnO nanostructures grown by metal-organic chemical vapor deposition. J. Mater. Sci. Mater. Electron. **19**(8), 760–763 (2007). https://doi.org/10.1007/s10854-007-9404-4

28. A. Rivera, J. Zeller, A. Sood, M. Anwar, A Comparison of ZnO Nanowires and Nanorods Grown Using MOCVD and Hydrothermal Processes. J. Electron. Mater. **42**(5), 894–900 (2013). https://doi.org/10.1007/s11664-012-2444-4
29. J.H. Liang, H.Y. Lai, Y.J. Chen, Morphology transition of ZnO films with DMZn flow rate in MOCVD process. Appl. Surf. Sci. **256**(23), 7305–7310 (2010). https://doi.org/10.1016/j.apsusc.2010.05.070
30. V. Sallet, F. Falyouni, A. Zeuner, A. Lusson, P. Galtier, Some Aspects of the MOCVD Growth of ZnO Nanorods by Using N_2O. J. Korean Phys. Soc. **53**(9), 3051–3054 (2008). https://doi.org/10.3938/jkps.53.3051
31. F. Falyouni, L. Benmamas, C. Thiandoume, J. Barjon, A. Lusson, P. Galtier, V. Sallet, Metal organic chemical vapor deposition growth and luminescence of ZnO micro- and nanowires. J. Vac. Sci. Technol. B **27**(3), 1662–1666 (2009). https://doi.org/10.1116/1.3137017
32. N. Thanh Son, J. -S. Noh, S. Park, Role of ZnO thin film in the vertically aligned growth of ZnO nanorods by chemical bath deposition, App. Surf. Sci. **379**, pp. 440–445 (2016). https://doi.org/10.1016/j.apsusc.2016.04.107
33. A.F. Abdulrahman, S.M. Ahmed, S.M. Hamad, A.A. Barzinjy, Effect of growth temperature on morphological, structural, and optical properties of ZnO Nanorods using modified chemical bath deposition method. J. Electron. Mater. **50**, 1482–1495 (2021). https://doi.org/10.1007/s11664-020-08705-7

Chapter 3
Mechanisms of ZnO Photocatalysis

The fundamental mechanism of the photocatalytic process involves the absorption of light photons by semiconductors, leading to the formation of electron (e^-) and hole (h^+) pairs in the valence band. When the ZnO photocatalyst is irradiated with light of energy greater than its optical band gap, electrons are excited from the valence band (VB) to the conduction band (CB), simultaneously creating an equal number of holes in the VB. These electron–hole pairs then migrate to the surface of ZnO, participating in oxidation and reduction reactions. Electrons react with oxygen to produce superoxide radical anions ($O_2^{\bullet-}$), while holes react with water to generate hydroxyl radicals (•OH). These hydroxyl radicals are primarily responsible for degrading pollutants into CO_2 and H_2O. Figure 3.1 illustrates the basic photodegradation mechanism of contaminants using ZnO.

Fig. 3.1 The basic mechanism of ZnO photocatalysis

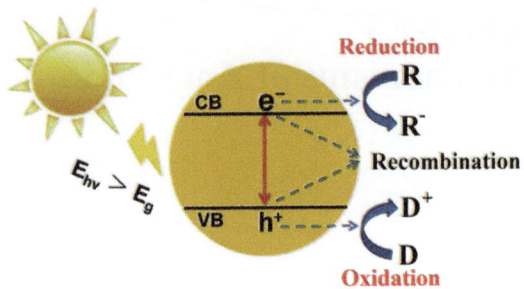

Chapter 4
Methods to Enhance ZnO Nanorods Photocatalytic Activity

ZnO nanorods face two major challenges that limit their photocatalytic efficiency: their limited absorption of UV light, which constitutes only a small portion of the solar spectrum, and a high recombination rate of electron–hole pairs, hindering charge separation. To address these challenges a diverse array of strategies has been employed. These include doping with metal, nonmetal, or rare earth elements; depositing noble metals; coupling with other semiconductors or carbon-based materials; and forming ternary nanocomposites. This section will review these methods.

4.1 ZnO Nanorods Doped with Metals or Nonmetals

A key method for enhancing the photocatalytic activity of ZnO nanorods is doping with metal or nonmetal atoms. This doping extends the material's spectral response into the visible light region by effectively reducing the optical band gap of ZnO NRs. Dopant atoms change the coordination environment of Zn atoms in ZnO, affecting its electronic structure by adding localized electronic levels within the optical band gap, raising the valence band maximum, or lowering the conduction band minimum.

4.1.1 ZnO Nanorods Doped with Transition Metals

One effective way to enhance the photocatalytic activity of ZnO nanorods is to dope them with transition metals such as Co [1–3], Cu [4], Mn [5], and Fe [6]. Transition metals with atomic radii similar to zinc can be easily incorporated into the ZnO lattice. Doping ZnO nanorods with these cations effectively improve charge separation and reduces the recombination rate of photoexcited electron–hole pairs, thereby enhancing photocatalytic activity. For example, Poornaprakash et al. [2] synthesized pristine and Co-doped ZnO nanorods using a facile hydrothermal method, as illustrated in Fig. 4.1a. Structural studies using X-ray diffraction (XRD) confirmed the incorporation of Co ions into the ZnO lattice. X-ray photoelectron spectroscopy (XPS) indicated that these Co ions were in the Co^{2+} oxidation state at Zn^{2+} sites. The reduction in optical band gap due to Co inclusion was assessed through diffuse reflectance spectroscopy and Kubelka–Munk plots for both pristine and Co-doped ZnO nanorods (Fig. 4.1b). The photocatalytic activity of these nanorods was evaluated by degrading Rhodamine-B under artificial solar light, the results shown in Fig. 4.1c showed improved degradation efficiency in the Co-doped sample. This enhancement was due to increased trapping sites as seen by photoluminescence (Fig. 4.1d), a larger surface area, and enhanced electron–hole pair separation. A proposed photocatalytic mechanism of the Co-doped ZnO sample is shown in Fig. 4.1e

Co-doping is a technique used to enhance the photocatalytic activity of ZnO nanorods, with examples in the literature including Al-Fe [7] and Mn-Cu [8] co-doped ZnO nanorods. Khalid et al. [7] synthesized ZnO nanorods co-doped with aluminum (Al) and iron (Fe) using the hydrothermal method. These samples were

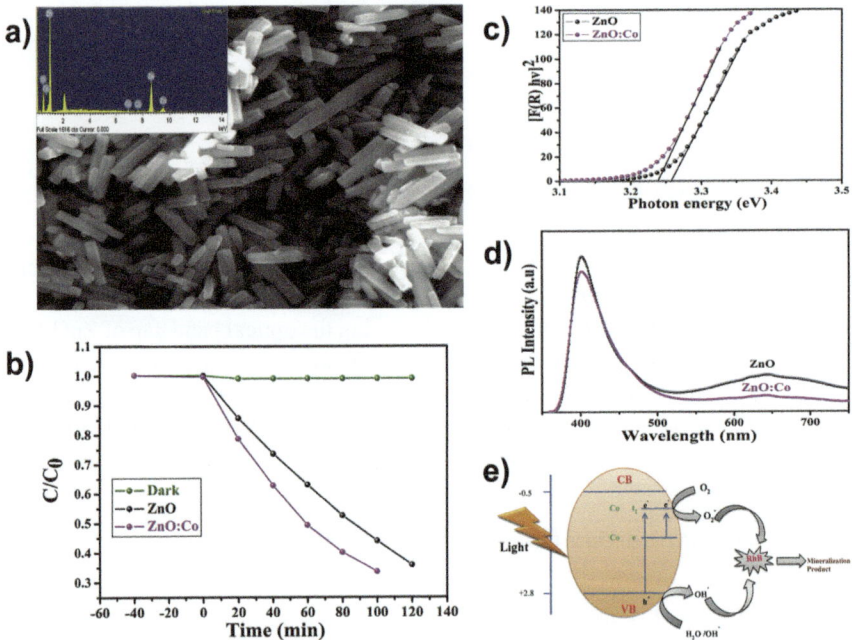

Fig. 4.1 **a** EDS (inset) with FESEM images of Co-doped ZnO nanorods. **b** Kubelka–Munk plots for pristine and Co-doped ZnO nanorods. **c** Time-dependent photocatalytic degradation ratios (C/C_0) of pristine and Co-doped nanorods **d** Photoluminescence spectra of pristine and Co-doped ZnO nanorods. **e** The schematic diagram for the photocatalytic mechanism of Co-doped ZnO sample [2]

4.1 ZnO Nanorods Doped with Metals or Nonmetals

tested as photocatalysts for degrading Methylene blue (MB) dye under visible light. Figure 4.2a shows the Tauc plot to get the optical band gap under direct transitions. Figure 4.2b shows that the Al-Fe/ZnO photocatalyst exhibited superior efficiency compared to Al-ZnO, Fe-ZnO, and bare ZnO nanorods. The enhanced activity of the Al-Fe/ZnO photocatalyst is attributed to the synergistic effects of Al and Fe dopants. The lifetime of charge carriers was significantly increased due to the incorporation of Al and Fe into the ZnO matrix, as confirmed by the PL spectrum shown in Fig. 4.2c. This extended lifetime of charge carriers is crucial for generating the radicals (superoxide and hydroxyl) needed to degrade toxic organic compounds. The PL analysis indicated that the inclusion of Al and Fe into the ZnO structure significantly reduced the recombination of electron–hole pairs. Additionally, absorption measurements confirmed a reduction in the ZnO optical band gap following the introduction of Al and Fe, due to the creation of new energy levels both above the valence band and below the conduction band, as shown in Fig. 4.2d.

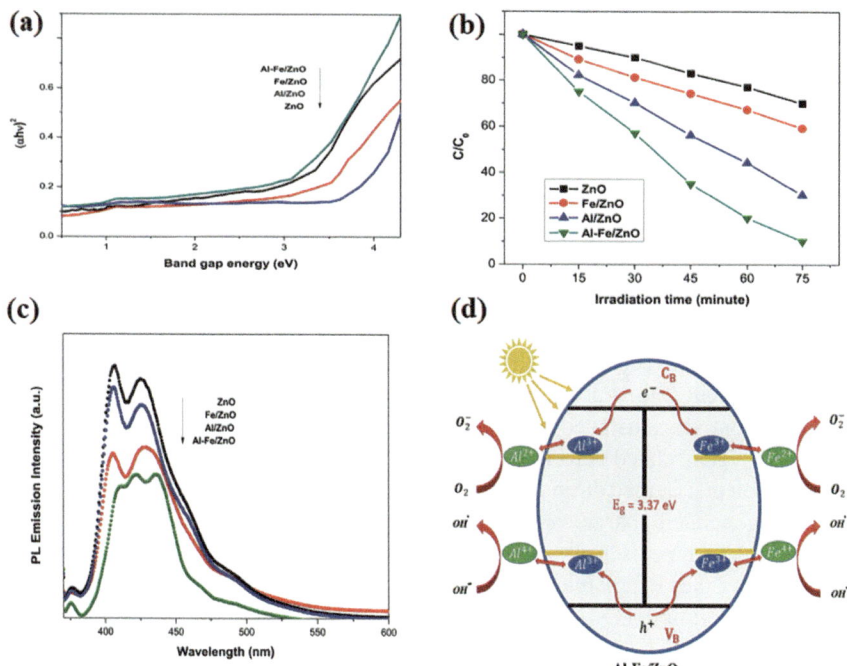

Fig. 4.2 a Tauc plot of different samples. b Photocatalytic test of different photocatalysts for MB dye degradation under visible-light illumination ($\lambda \geq 400$ nm). c PL spectra of ZnO, Fe/ZnO, AL/ZnO, and Al-Fe/ZnO nanorods photocatalyst. d Schematic diagram of the proposed reaction mechanism for MB dye degradation using Al-Fe/ZnO photocatalyst [7]

4.1.2 ZnO Nanorods Doped with Rare Earth Metals

Doping ZnO nanorods (NRs) with rare-earth (RE) metals induces localized energy levels within the ZnO band structure, which modifies its electronic properties and enhances its visible light absorption capacity. This phenomenon occurs due to the interaction between the charge transfer processes of ZnO's valence or conduction bands and the unique 4f or 5d orbitals of the RE ions. The inclusion of RE ions, such as Sm, Ce, Eu, La, and Gd, substantially alters the electronic structure of ZnO by introducing oxygen-related defects, enhancing the separation and lifetimes of photogenerated charge carriers, and ultimately increasing the material's photocatalytic efficiency. However, incorporating RE ions into the ZnO lattice presents significant challenges, primarily due to two key factors. The first challenge is the low saturation concentration of RE ions in the ZnO matrix, which stems from the considerable differences in ionic radii and charge between RE^{3+} ions and Zn^{2+}. This size and charge disparity limits the solubility of RE ions within the ZnO lattice, often leading to the formation of secondary phases or defects. The second challenge arises from the energy level mismatch between the RE ions and the ZnO valence and conduction bands, particularly since the charge transfer level of RE^{3+} ions is located at or above the conduction band minimum of ZnO, as reported by Khatamian et al. [9]. This positioning hinders efficient energy transfer between ZnO and the RE^{3+} ions, impacting their overall contribution to the material's photocatalytic activity. For example, Ranjith Kumar et al. [10] synthesized vertically aligned Sm-doped ZnO nanorod arrays using a simple vapor transport method on silicon substrates. Their study demonstrated that Sm doping resulted in shorter, misaligned nanorods (Fig. 4.3a) but significantly suppressed electron–hole recombination, as evidenced by photoluminescence (PL) analysis (Fig. 4.3b). The Sm-doped ZnO NRs exhibited enhanced photocatalytic efficiency for methylene blue (MB) degradation under natural sunlight (Fig. 4.3c), surpassing the performance of undoped ZnO NRs. The improvement was attributed to Sm impurities, which enhanced electron transfer and created oxygen-related defects that altered the ZnO band structure (Fig. 4.3d). These oxygen defects, formed during doping, played a crucial role in prolonging the lifetime of photogenerated charge carriers, thereby improving photocatalytic performance. Similar enhancements in photocatalytic activity have been observed in ZnO nanorods doped with other RE elements such as cerium [11], europium [12], lanthanum [13, 14], and gadolinium [15].

4.1.3 Deposition of Noble Metals Onto ZnO Nanorods

Deposition of noble metal nanoparticles (NPs) such as Ag [16, 17], Au [18–20], and Pt [21–24] on the surface of ZnO Nanorods improves photocatalytic performance by promoting charge transfer and thus decreasing electron–hole pair recombination rate. For instance, the increase in photoactivity can be explained by the mechanism

4.1 ZnO Nanorods Doped with Metals or Nonmetals

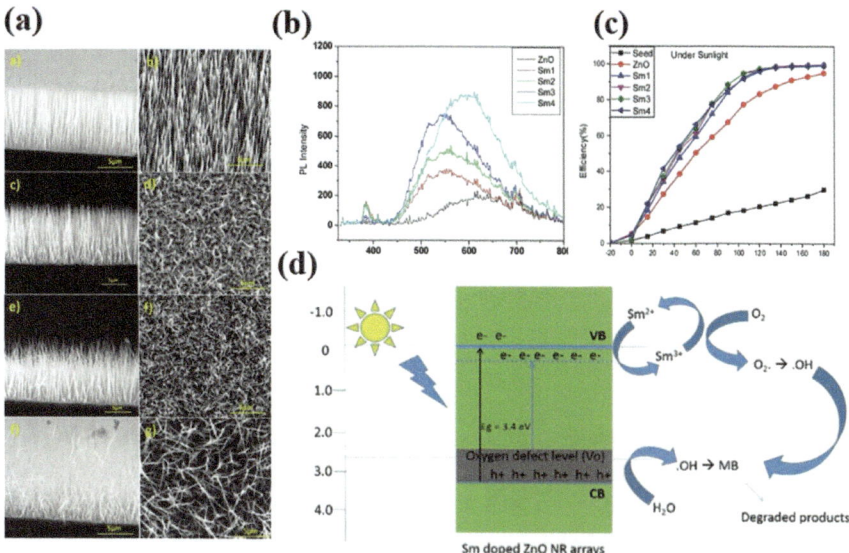

Fig. 4.3 a SEM image of top and cross-section view of Sm doped ZnO NR arrays (a, b) undoped ZnO, (c, d) 2 wt.% Sm doped ZnO, (e, f) 4 wt.% Sm doped ZnO, and (g, h) 6 wt.% Sm doped ZnO. **b** Room Temperature PL spectra of the different concentrations of Sm doped ZnO NR arrays. **c** Photodegradation properties of pure and Sm doped ZnO NRs array under visible light irradiation. **d** Band structure of Sm-doped ZnO NR arrays as photocatalysts under illumination [10]

depicted in Fig. 4.4, which summarizes the different processes that occur during the photocatalytic degradation of MO under UV or visible light illumination. Figure 4.4a shows the charge transfer process under UV light irradiation over Au-ZnO NRs, where an excited electron in the valence band (VB) is excited into the conduction band (CB) while simultaneously leaving a hole (h^+) in the VB. Since the work function of ZnO (4.45 eV) is lower than that of AuNPs (5.31 eV), the Fermi level of AuNPs is below the CB of ZnO, and photo-excited electrons are transferred from the CB of ZnO to the AuNPs. This occurs because the presence of metallic noble NPs such as Au, Ag, or Pt deposited on a semiconductor contributes to forming a Schottky barrier that serves as a container for the Photoexcited electrons. As a result, direct contact of AuNPs with ZnO enhances charge carrier separation and thus contributes to charge carrier recombination reduction and improves photocatalytic activity. Figure 4.4b shows the charge transfer process under visible light irradiation over Au-ZnO NRs, and the increase of photoactivity can be explained as follows. ZnO is a semiconductor with a wide optical band gap of 3.1 eV, which can only absorb UV light. Thus, the ZnO cannot be excited by visible light. Nevertheless, metallic NPs have good absorption in the visible light region of the electromagnetic spectrum due to their special surface plasmon resonance (SPR). SPR-induced electrons produced under visible radiation on the AuNPs can migrate to the CB of ZnO NRs through the metal-ZnO NRs interface and then participate in the photocatalytic process, increasing the photocatalytic activity of ZnO NRs.

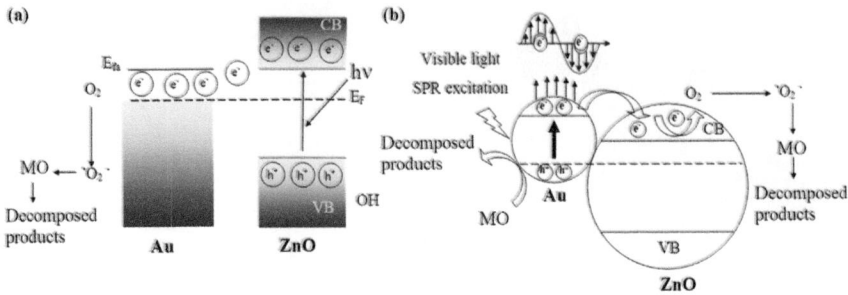

Fig. 4.4 Schematic band diagram of noble metal-ZnO NRs showing the charge transfer process under **a** UV light irradiation and **b** visible light irradiation [25]

On the other hand, Nana et al. [18] used a biogenic method to fabricate gold-anchored ZnO nanorods (Au-ZnO NRs) as shown in Fig. 4.5a. PL spectra in Fig. 4.5b show that the deposition of Au NPs reduces the photoluminescence emission from Au-ZnO NRs, indicating enhanced separation and transfer of photogenerated charge carriers. The fabricated samples have been evaluated in the photodegradation of methyl orange (MO) under UV and sunlight irradiation as seen in Fig. 4.5c, and the Au-ZnO NRs exhibited a significantly enhanced photocatalytic activity compared to pure ZnO NRs, as shown in Fig. 4.5d.

4.1.4 ZnO Nanorods Doped with Nonmetals

The use of nonmetals such as nitrogen (N) [26–29], carbon (C) [30, 31], and sulfur (S) [32–35] as dopant elements in ZnO Nanorods cause a modification of the conduction band to more negative potentials, which results in a narrowing of the optical band gap and thus an increased absorption at lower energies in the solar spectrum. N has been considered one of the most suitable nonmetal elements for p-type dopants due to its nontoxicity, similar atomic radius, and lower electronegativity and ionization energy than the O atom. Due to N and O's nearly identical ionic radius, mixing its 2p orbital with its 2p orbital of O contributes significantly to bandgap narrowing, improving photocatalytic activity [27]. Figure 4.6a shows N-doped ZnO NRs synthesized by Meng Wang et al. [27]. FESEM images of pure ZnO are shown in Fig. 4.6b. The N elements were incorporated into the lattice of hydrothermally grown ZnO Nanorods using an ion implantation method. N doping did not change the rod-like morphology of ZnO NRs but extended the optical absorption edges to the visible light region, as shown in Fig. 4.6c. Additionally, the gradient N dopant distribution resulted in a terraced band structure and thus a built-in electric field, contributing to the enhanced separation efficiency of photogenerated electron–hole pairs in ZnO NRs.

On the other hand, phosphorus-doped ZnO (P-ZnO) nanostructures have been increasingly drawing concern in applications in photocatalytic water splitting [36,

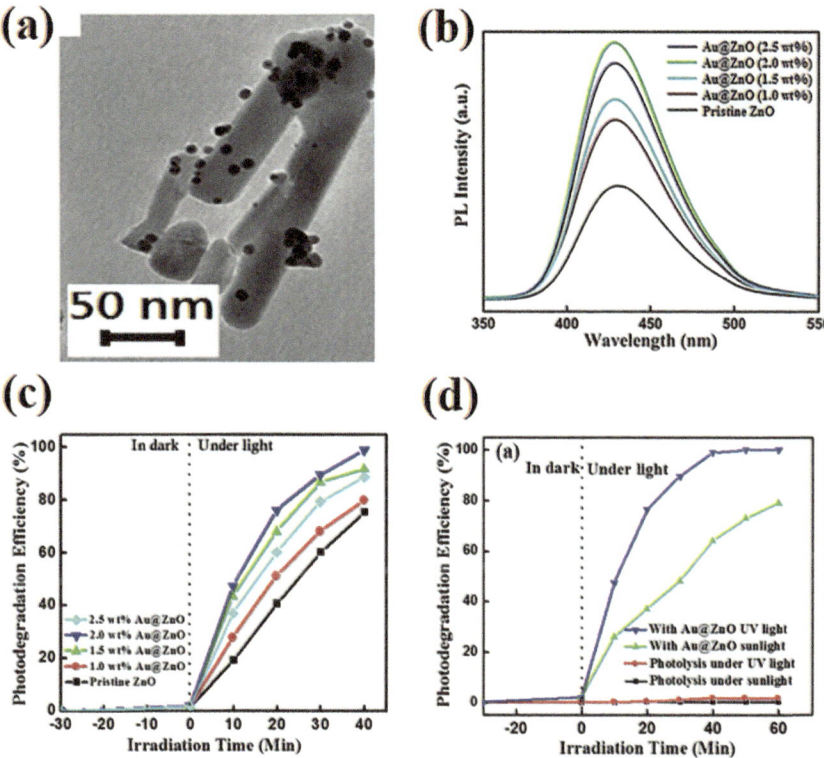

Fig. 4.5 a TEM image of Au-ZnO (2.0 wt. %). b PL spectral changes of the Au@ZnO with different gold contents. c Effects of gold content on the photodegradation of MO c) kinetics of MO degradation under UV light. d Effects of the light source (UV light and sunlight) on the degradation of MO [18]

37]. The reported studies demonstrate that phosphorus doping efficiently promotes carrier transfer and changes the band structure by narrowing down the bandgap energy, which helps to expand the light adsorption of ZnO to visible light increasing the conversion efficiency of photocatalytic water splitting [38]. In addition, doping provides electron trap states and inhibits electron–hole recombination [38]. However, despite these advantages, there is limited information on using P-doped ZnO nanostructures for wastewater treatment through photocatalytic processes and even less on using their nanorod morphology as photocatalysts. This lack of information is due to the great challenge of obtaining p-type ZnO with high photocatalytic activity and stability, which has become a significant obstacle to the wide application of ZnO, although many strategies are currently being proposed and implemented [38, 39]. Therefore, increasing focus on constructing p-type ZnO is highly desirable, as it could offer more opportunities to enhance photocatalytic water treatment systems.

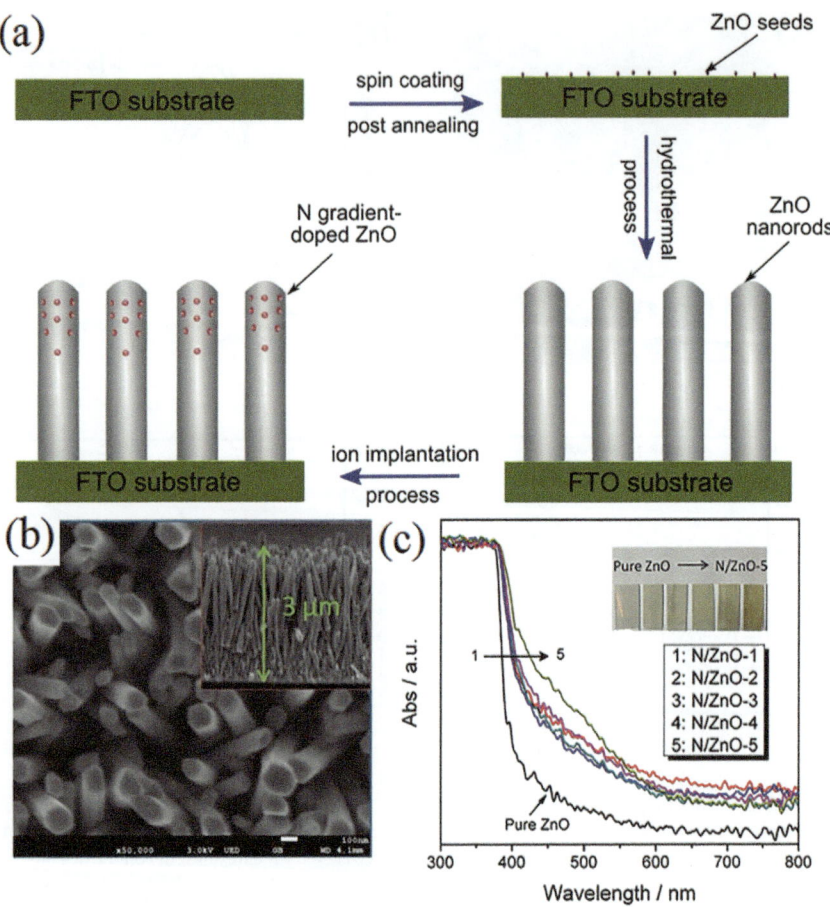

Fig. 4.6 a Schematic diagram of the preparation process for N gradient-doped ZnO NRs using the advanced ion implantation method. b FESEM images of pure ZnO. c UV–Vis spectra and digital photograph (inset) of N-doped ZnO NRs with pristine ZnO as the reference [27]

4.2 Heterojunctions

Coupling ZnO with narrow optical band gap semiconductors such as CuO [40], CdS [41], and ZnSe [42] enhances charge separation and improves the sunlight absorption of ZnO Nanorods, thereby improving their photocatalytic performance. For instance, Yuanyuan et al. [40] synthesized CuO nanoparticle-loaded ZnO nanowire arrays (CuO-ZnO NWAs) on ITO glass using a three-step chemical approach. Morphology analysis indicates that CuO nanoparticles are dispersed on the surface of the ZnO NWAs, as seen in Fig. 4.7a with the FESEM images showing the top view, cross-sectional, and Energy-dispersive X-ray spectra of CuO-ZnO NWAs. Compared to ZnO NWAs, the CuO-ZnO NWAs have an absorption peak that extends into the

4.2 Heterojunctions

visible region, as shown in Fig. 4.7b. Additionally, CuO-ZnO NWAs exhibit a higher photocurrent response and photodegradation efficiency of methylene blue (MB) under visible light irradiation than pure ZnO NWAs, as clearly shown in Fig. 4.7c, which can be attributed to the enhanced efficiency and broadened absorption range of visible light, as well as the efficient generation and separation of photogenerated electrons and holes due to the formation of CuO-ZnO (p-n) heterogeneous structure. Figure 4.7d shows a scheme of the formation and mechanism of the junction in photocatalysis. According to the plot, the charge carriers migrate in the following manner: Without irradiation, the electrons in ZnO NWAs migrate to CuO, while the holes in CuO migrate to ZnO NWAs to achieve the Fermi level equilibrium for the heterostructure, resulting in the build-up of the internal electric field. However, light irradiation excites both n-type ZnO NWAs and p-type CuO, leading to the generation of charge carrier pairs. They will then be effectively separated and transferred under the influence of the internal electric field. Due to the band locations, this step is also thermodynamically suitable. Thus, the synergistic effect of the internal electric field and band alignment of the p-n heterojunction improve the photocatalytic efficiency of CuO-ZnO NWAs.

Gholamveysi et al. [43] investigated the photocatalytic degradation of the tetracycline (TC) drug and the dye Rhodamine B (RhB) using ZnO Nanorods, $MoSe_2$, and ZnO Nanorods/$MoSe_2$ nanocomposites under Xe light illumination, with the results displayed in Fig. 4.8. The degradation activity of TC and RhB over time is compared in Figs. 4.8a, b, respectively. The removal efficiencies after 105 min of irradiation were 39.3% with ZnO Nanorods and 25.2% with $MoSe_2$ for TC, and 28.3% and 25.8% for RhB with ZnO Nanorods and $MoSe_2$, respectively. However, the ZnO Nanorods/$MoSe_2$ nanocomposite significantly outperformed both ZnO Nanorods and $MoSe_2$, achieving removal efficiencies of 70.2% for TC and 62.3% for RhB. This confirms the enhanced degradation facilitated by the ZnO Nanorods/$MoSe_2$ binary composite. Figs. 4.8c, d show ln (C_0/C) over time, indicating that the degradation followed pseudo-first-order kinetics, with good R-squared values (> 0.99) in all cases. The apparent rate constant (k) was determined by fitting the curve and calculating the slope of the plot of $ln(C_0/C)$ versus time. For TC removal (Fig. 4.8c), the k values increased by approximately 2.2 and 5.1 times, from 59.9×10^{-4} min^{-1} for ZnO Nanorods and 25.9×10^{-4} min^{-1} for $MoSe_2$ to 130.8×10^{-4} min^{-1} for the ZnO Nanorods/$MoSe_2$ binary composite. Additionally, the photocatalytic RhB degradation rate constant (Fig. 4.8c) for the ZnO Nanorods/$MoSe_2$ nanocomposite was 114.3×10^{-4} min^{-1}, about 3.1 and 3.7 times higher than that of ZnO Nanorods (37.4×10^{-4} min^{-1}) and $MoSe_2$ (31.2×10^{-4} min^{-1}), respectively. By comparing the degradation rate constants, it is confirmed that the degradation of TC and RhB contaminants was most rapid with the ZnO Nanorods/$MoSe_2$ sample. This enhancement in degradation can be attributed to the advantageous coupling of ZnO and $MoSe_2$, forming a binary composite that benefits from the synergistic effect in the photocatalysis process by improving the effective separation of photo-generated electron–hole pairs [43].

Traditional mechanisms, such as the conventional type II heterojunction model, are unsuitable for photocatalytic reactions. While this model explains the spatial

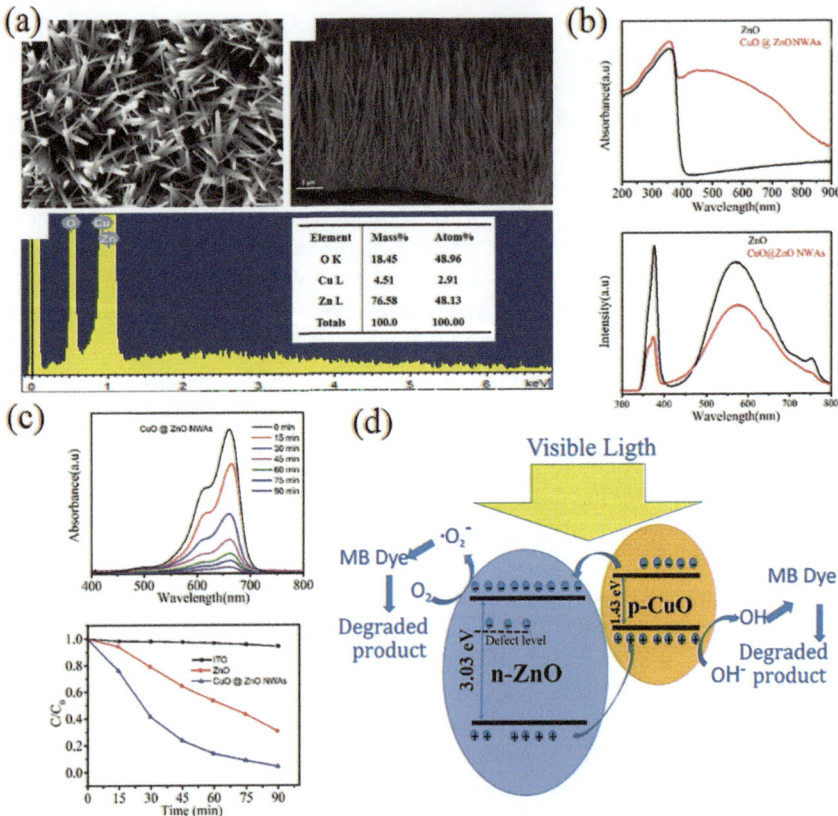

Fig. 4.7 a FESEM images of the top view, cross-sectional, and Energy-dispersive X-ray spectra of CuO-ZnO NWAs. **b** UV–vis diffuse reflectance spectra and PL of ZnO NWAs and CuO-ZnO NWAs. **c** Absorption spectra of MB aqueous solution of the as-synthesized CuO-ZnO NWAs under visible light irradiation and Photodegradation efficiency over different photocatalysts. **d** Schematic photocatalytic mechanism of the p–n heterojunctions of CuO-ZnO NWAs [40]

separation of photo-induced charges, limiting their redox capability, i.e. electrons accumulate at the conduction band (CB) minimum and holes at the valence band (VB) maximum [44]. Having this in mind, a proposed mechanism of the photo-electrocatalysis system using ZnO Nanorods/MoSe$_2$/fluorine-doped tin oxide (ZnO/MoSe$_2$/FTO) samples as photoanodes is presented in Fig. 4.9, in which a new step-scheme (S-scheme) heterojunction mechanism has been proposed, which involves two n-type semiconductor photocatalysts. In this setup, MoSe$_2$ has higher conduction band and valence band energies, as well as a higher Fermi energy (E_F), compared to ZnO Nanorods. Due to this difference, electrons from E_F of MoSe$_2$ naturally migrate to E_F of ZnO Nanorods until equilibrium is achieved at the interface. This electron transfer leaves a negative charge on the ZnO Nanorods side and a positive charge on the MoSe$_2$ side, creating an internal electric field that drives the charge transfer

4.2 Heterojunctions

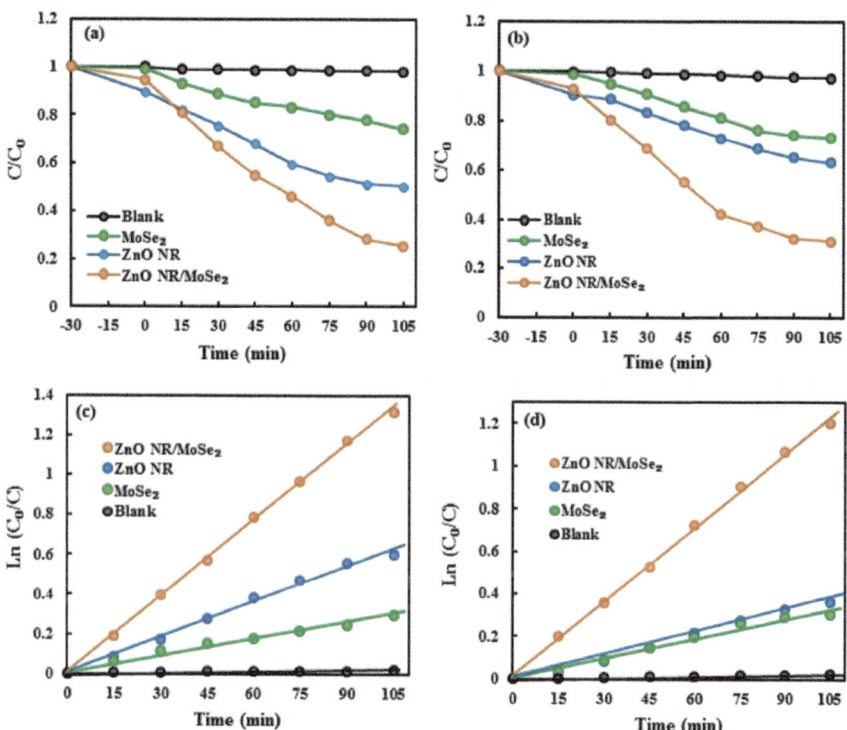

Fig. 4.8 **a, b** Photocatalytic degradation curves and **c, d** the kinetic curves for TC and RhB degradation with different photoelectrocatalysts [43]

process. This results in an electron depletion region on the $MoSe_2$ side, causing an upward bending of its CB and VB, and an electron accumulation region on the ZnO Nanorods side, resulting in a downward bending of its CB and VB. When the system was irradiated, $MoSe_2$ and ZnO Nanorods generated electrons and holes. The internal electric field, band bending, and electrostatic attraction between electrons in the CB of ZnO Nanorods and holes in the VB of $MoSe_2$ facilitate the recombination of these charges. In $MoSe_2$, the remaining electrons in the CB can reduce O_2 molecules to produce •O_2 radicals when a sufficiently negative potential is applied. Meanwhile, the holes in the VB of ZnO Nanorods can oxidize H_2O or OH^- molecules to generate •OH radicals. This process maximizes the system's reduction and oxidation capabilities. Additionally, some electrons in the ZnO Nanorods/$MoSe_2$/FTO photoanode can migrate to the Pt cathode due to an externally applied potential. The Pt cathode catalyzes the absorption of O_2 to generate •O_2^- species, which contribute to degradation activities. Thus, the S-scheme mechanism in the ZnO Nanorods/$MoSe_2$/FTO photoanode, combined with the electrochemical process, explains the enhances the photoelectrocatalytic performance for the degradation of TC and RhB.

Coupling ZnO Nanorods with another semiconductor of the same type, with properly aligned band positions, where the conduction band of one and the valence

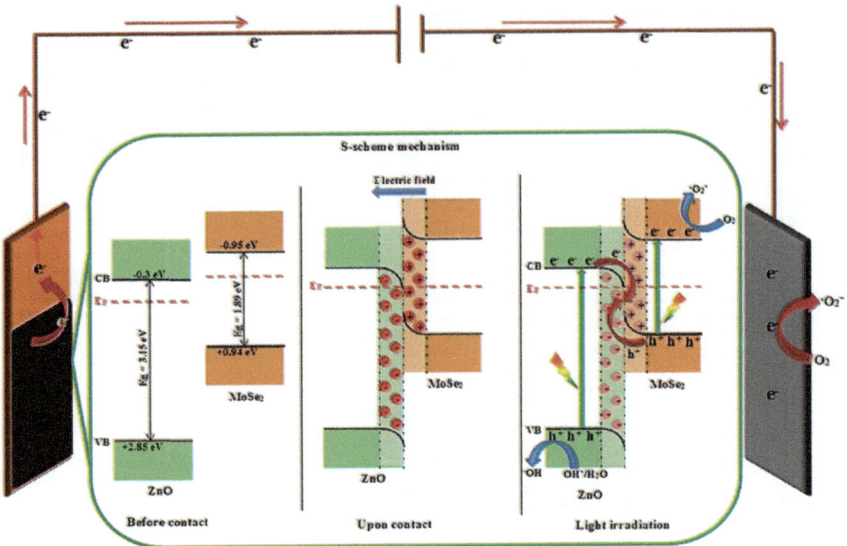

Fig. 4.9 Illustrating the photoelectrocatalytic mechanism for pollutant degradation at the interface of ZnO Nanorods and MoSe$_2$, including the roles of the internal electric field, band edge bending, and S-scheme charge transfer mechanism, as well as charge transfer in the external circuit [43]

band of the other enable the transfer of electron–hole pairs, can improve performance, as shown in Fig. 4.10, could lead to enhanced photocatalytic performance compared with the pristine ZnO Nanorods. For instance, Ramos et al. [45] fabricated ZnO/TiO$_2$ nanostructures via an electrostatically modified electrospinning technique onto fluorine-doped tin oxide (FTO) substrate, and the photocatalytic activity was evaluated towards the photodegradation of methyl orange. The results show that coupling ZnO with TiO$_2$, both n-type semiconductors, in an optimal ratio can enhance photocatalytic performance compared to pure ZnO. This improvement is primarily attributed to effective charge separation at the interface. The mechanism of charge separation and photocatalytic reaction for ZnO/TiO$_2$ nanostructures was proposed and represented in Fig. 4.10. As illustrated in the scheme, when the ZnO/TiO$_2$ heterojunction is irradiated from the light source, with photon energy higher or equal to the optical band gap energy of ZnO and TiO$_2$, the electrons (e$^-$) are excited from the valence band (VB) to the conduction band (CB) with simultaneous generation of the same number of holes (h$^+$) in the VB. The electron transfer occurs from the CB of ZnO to the CB of TiO$_2$, and conversely, the hole transfer takes place from the VB of TiO$_2$ to the VB of ZnO. Moreover, due to the synergistic effects between ZnO and TiO$_2$, the recombination rate of the photogenerated electrons and holes was suppressed, and the lifetime of the photogenerated charge carriers was increased in the composite nanostructures. Consequently, this research demonstrated an enhancement in the photocatalytic activity of the ZnO/TiO$_2$ heterojunction.

4.3 ZnO Nanorods-Based Carbonaceous Materials

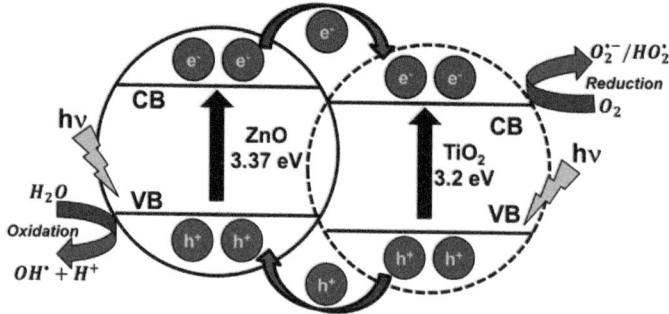

Fig. 4.10 A schematic diagram for the charge-transfer and photocatalytic process of ZnO/TiO$_2$ nanofibers for methyl orange degradation [45]

4.3 ZnO Nanorods-Based Carbonaceous Materials

Carbonaceous materials have attracted significant research attention worldwide over the past few decades [46, 47]. Various carbon materials, including activated carbon, carbon nanotubes, graphene-based materials, and carbon dots, exhibit unique properties such as high electron mobility, excellent adsorption, conductivity, tunable chemical characteristics, and a large delocalized π-electron system [47–49]. These properties have made them popular in the design of photocatalyst systems. When coupled with ZnO Nanorods, carbonaceous materials significantly enhance photocatalytic activity by acting as efficient electron acceptors, facilitating charge transfer from ZnO, and preventing electron–hole pair recombination [50–52]. This enhancement improves the photodegradation of various pollutants under visible light.

The ZnO/graphene Nanorods have been synthesized using various methods and on diverse substrates, including SiO$_2$ and Si wafers, fluorine-doped tin oxide (FTO), indium tin oxide (ITO) glasses, and nickel foam, among others [52–54]. There are diversified ways to include graphene in the ZnO Nanorods fabrication process; previous research reported the growth of ZnO Nanorod arrays on graphene or ZnO/graphene-based seed layers [55–57], while in other studies, graphene and reduced graphene oxide (rGO) sheets were aggregated on the surface of ZnO Nanorods [58–60].

Zou et al. reported the ZnO/graphene Nanorods fabrication on glass substrates [60]. In their procedure, a homogeneous suspension of reduced graphene sheets (rGss) was first prepared using a modified Hummers method. This suspension was then spin-coated onto glass substrates, which were subsequently annealed under a nitrogen (N$_2$) flow to ensure a firm coating. Subsequently, for the growth of ZnO/graphene Nanorods, a zinc oxide precursor solution was sprayed on the rGss, and then it was dried and annealed to form ZnO nanocrystals seeded on the rGss. After that, the substrates seeded with the ZnO/rGss composites were put in a growth solution consisting of zinc nitrate hexahydrate, hexamethylenetetramine, and deionized water to form the ZnO Nanorods arrays on a single side of the flexible rGss forming two-layered heterostructures of ZnO/graphene.

Moreover, Ramos et al. [61, 62] fabricated ZnO and ZnO-rGO Nanorods onto FTO glass plates by a wet chemical method from ZnO and ZnO-rGO seeds layers fabricated by electrospinning, respectively. A spinning solution was prepared by dissolving zinc acetate and polyvinylpyrrolidone in N, N-dimethylformamide to synthesize a pure ZnO seed layer. Two diverse series of ZnO-rGO seed layers were obtained under different conditions. The first series of ZnO-rGO seed layers were fabricated from the ZnO spinning solution adding 0.1, 0.2, and 0.3 wt.% rGO. In the second series, a spinning solution containing 0.2 wt.% rGO was used, and three different spinning voltages were applied: 20, 30, and 40 kV. Then, the ZnO-rGO seed layers were obtained by calcination in a muffle furnace at 400 °C of the coated substrates. The solution medium employed for the growth of the ZnO and ZnO-rGO NRs and the steps performed to obtain them were carried out based on Rodriguez et al. [63]. Figure 4.11 illustrates the formation procedure of ZnO-rGO Nanorod arrays.

On the other hand, Rokhsat and Akhavan deposited approximately 1 wt.% of graphene oxide (GO) sheets on the surface of ZnO Nanorods films, which were obtained previously from the hydrothermal growth of zinc oxide seed layers synthesized by spin coating [59]. ZnO Nanorods were deposited onto glass substrates and then coated with a layer of graphene oxide (GO) in a two-step process. First, the GO was applied using the drop-casting method, followed by spin coating to ensure uniform coverage. Finally, to obtain a chemical bonding between the ZnO and GO

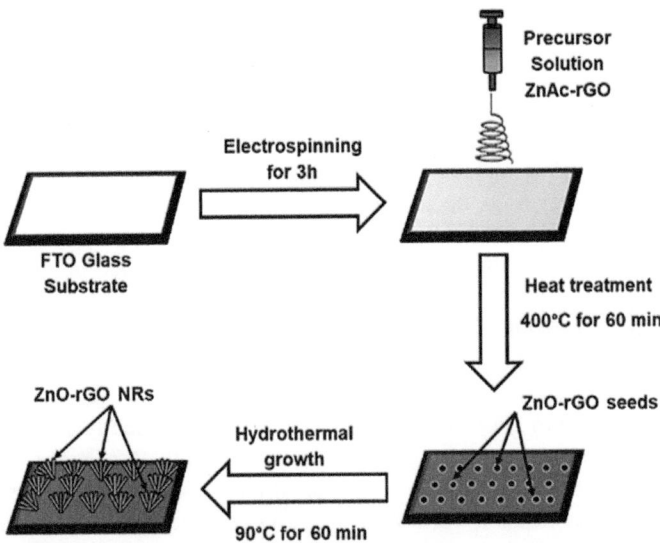

Fig. 4.11 Schematic diagram illustrating the process for producing ZnO-rGO Nanorod arrays from ZnO and ZnO-rGO seed layers fabricated by electrospinning [62]

4.3 ZnO Nanorods-Based Carbonaceous Materials

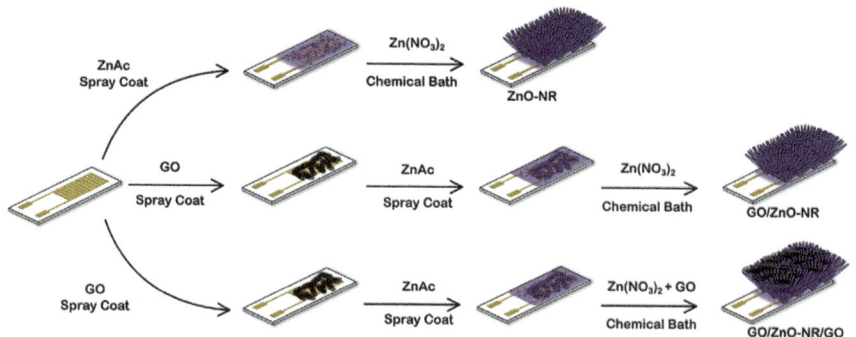

Fig. 4.12 Scheme illustrating the fabrication of ZnO-NRs, GO/ZnO-NRs, and GO/ZnO-NRs/GO samples [53]

sheets, the spin-coated films obtained were annealed, thus forming the GO/ZnO Nanorods films.

Furthermore, it is also possible to obtain by the same fabrication method composites with graphene oxide sheets under ZnO Nanorods (GO/ZnO-NRs), or ZnO Nanorods between GO sheets (GO/ZnO-NRs/GO) when GO is placed in the bath during the growth of GO/ZnO NRs as shown by the research done by Vessalli et al. [53]. In that work, the GO/ZnO and GO/ZnO/GO Nanorods were grown via chemical bath deposition (CBD) on alumina substrates. The seed layers were obtained from the sprayed GO layers, followed by depositing a zinc acetate on the substrate. After that, the Nanorods were grown by CBD. For GO/ZnO Nanorods, a growth solution composed of hexamethylenetetramine and zinc nitrate was used, while for GO/ZnO NRs/GO synthesis a GO solution was added to the growth solution described above. Then, the hydrothermal growth of seed layers was performed, aiming to promote the ZnO NRs/GO and GO/Zn NRs/GO composites. Figure 4.12 shows an overview of the fabrication method.

Recently, graphene quantum dots (GQDs) have received much attention because they can improve the photoactive performances of ZnO Nanorods due to their size-tunable optical absorption and emission similar to semiconductor QDs [64, 65]. GQDs also have several advantages over semiconductor QDs, including their nontoxic nature, capability for large-scale production, chemical stability, strong optical absorption across a broad spectral range, and ease of synthesis [66, 67].

In previous research, Rahimi et al. [68, 69] obtained ZnO Nanorods/GQDs thin films onto glass substrates, and the ZnO Nanorods were fabricated via a solvothermal method as a first step. The GQDs were loaded onto the ZnO NRs thin films in two ways. In the first procedure, GQDs were successfully coated onto the synthesized ZnO Nanorods (NRs) using a dip-coating process for varying durations [70]. In the second procedure, GQDs were effectively deposited onto the thin films of ZnO NRs through a drop-casting method [68]. Finally, the substrates were annealed to significantly enhance the bonding between the ZnO NRs and the GQDs.

Alternatively, Kathalingam et al. [70] presented an easy and innovative method for creating vertically aligned ZnO NRs arrays wrapped in GQDs. The ZnO NRs/GQD heterostructure was created as a core/shell structure on an ITO substrate using a novel combination of three techniques: spin-coating, electron beam evaporation, and rapid thermal processing. In the first step, ZnO Nanorods deposited onto ITO substrates were obtained from growth by a hydrothermal process of ZnO seed layers which were fabricated by spin-coating. For the preparation of the GQD-decorated ZnO NRs, the ZnO NRs fabricated previously were coated with PMMA (poly(methyl methacrylate)); after that, the PMMA-coated ZnO NRs were annealed at high temperature to convert the PMMA to GQDs. For the thermal conversion of PMMA into GQD and to avoid evaporation, a metal-capping layer was used using an E-beam evaporator. Figure 4.13 illustrates the schematic diagram of the fabrication process for the ZnO Nanorods (NRs) and GQD heterostructure.

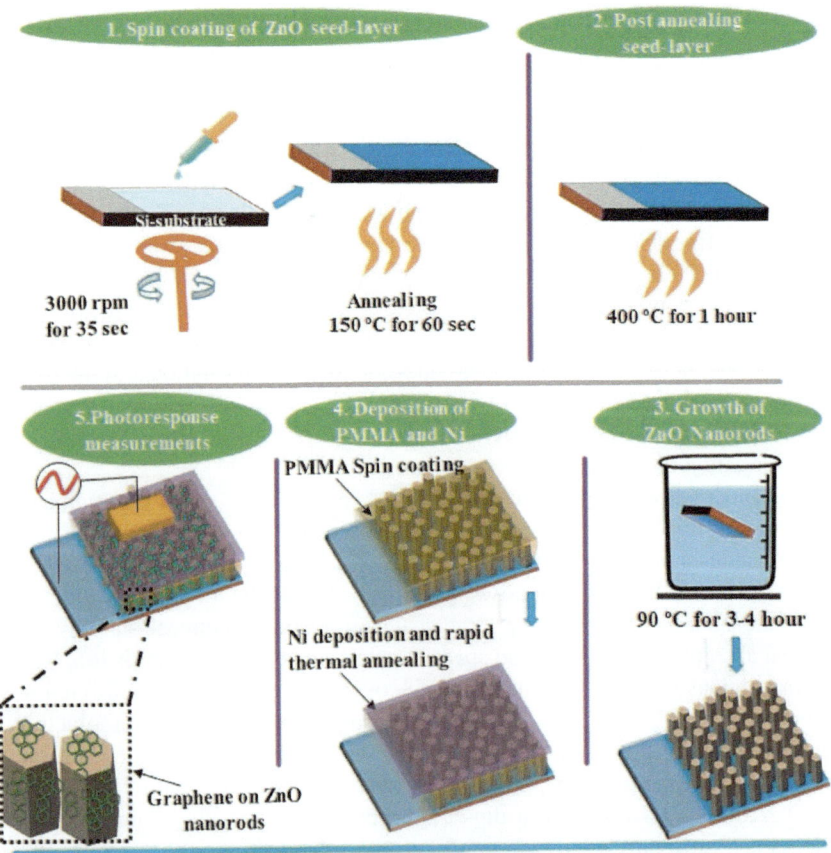

Fig. 4.13 A schematic of the ZnO Nanorods with GQDs fabrication process [70]

4.3 ZnO Nanorods-Based Carbonaceous Materials

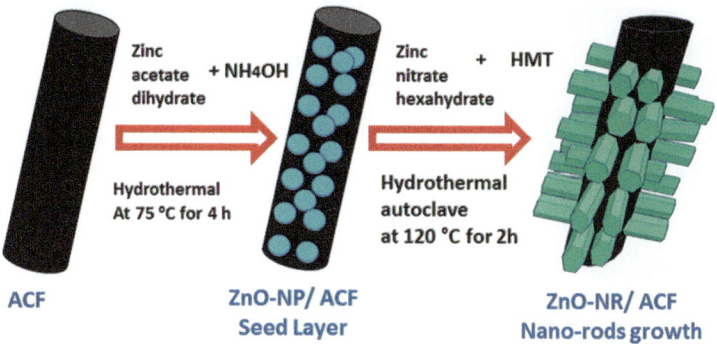

Fig. 4.14 Schematic diagram illustrating the synthesis process of zinc oxide Nanorods on activated carbon fibers (ZnO-NR/ACF) [71]

Albiss and Abu-Dalo [71] used the hydrothermal method to fabricate ZnO Nanorods on activated carbon fibers (ZnO-NR/ACF). ZnO nanoparticles were synthesized by chemical precipitation and deposited as seeds on porous activated carbon fiber substrates (ACF). The average fiber length was about 3 mm, with an average fiber diameter of 10 μm and a specific surface area of 1000 m^2/g, as stated by the manufacturer. The ZnO nanoparticles with the ACF nanocomposite were synthesized by mixing the ZnO nanoparticles and ACF through stirring. The obtained nanocomposite was filtered, washed several times with distilled water and ethanol, dried, and annealed to obtain the ZnO/ACF seed layer. Finally, zinc oxide Nanorods (ZnO-NR) were successfully grown on the seeds and assembled on the fiber surface in various patterns to form ZnO-NR/ACF nanocomposites. The obtained ZnO-NR/ACF nanocomposite was filtered, flushed with DI water until the pH of the final solution was 7.0, and dried in a vacuum oven. The amount of ZnO Nanorods was determined by measuring the weight difference of the ACFs before and after the hydrothermal process. Figure 4.14 shows an overview of the fabrication of the nanocomposites.

4.3.1 Application of ZnO Nanorods-Based Carbonaceous Photocatalyst for Dye Degradation

Many researchers have extensively studied the hybridization of carbon materials with ZnO Nanorods, emphasizing their crucial role in enhancing ZnO's photocatalytic performance due to their unique properties [72, 73]. Additionally, utilizing different substrates for depositing ZnO/carbon material Nanorods improves the ability to reuse the photocatalyst and facilitates the implementation of large-scale water purification systems.

In the work presented by Rokhsat and Akhavan ZnO/GO Nanorods were fabricated by depositing GO sheets on the surface of hydrothermally synthesized ZnO

Nanorods films [59]. Besides, some ZnO/GO samples were exposed to 400W UV irradiation for 1 h to finally investigate the photocatalytic activity of all nanocomposite films obtained based on methylene blue (MB) dye degradation under UV light. The optical band gap values estimated for ZnO, ZnO/GO, and UV-treated ZnO/GO Nanorods were approximately 3.78, 3.24, and 3.14 eV, respectively. These values were obtained from the curves of the optical absorption $(\alpha h\nu)^2$ versus incident photon energy $(h\nu)$ shown in Fig. 4.15a, based on the following relation [7]: $h = A(h\nu - E_g)^n$, where h, ν, A, and E_g are Planck's constant, frequency of light, proportionality constant, and optical band gap energy, respectively. The constant (n) value is 0.5 for the direct transition between the valence and conduction bands. The results obtained in this work show a significant effect on the optical properties of ZnO Nanorods due to the incorporation of graphene oxide and UV light treatment. To investigate the separation process of e^- and h^+ pairs, photoluminescence (PL) measurements were carried out. Fig. 4.15b shows the PL spectra of the samples. The results indicated that PL intensity significantly decreased for ZnO/GO Nanorods, especially the UV-treated ones, compared with PL intensity obtained for pure ZnO Nanorods. This quenching of the main emission peak can be ascribed to higher charge transfer at the interface from the ZnO Nanorods into GO sheets [74, 75]. The ZnO/GO Nanorods thin films, especially the UV-treated ones, could degrade 90% MB, with an initial concentration of 3 M, compared to the pure ZnO Nanorods, which only degraded 75% MB after 450 min UV irradiation (Fig. 4.15c). The improvement was assigned to the effective separation of the photo-excited electron–hole pairs between GO and ZnO NRs, as confirmed by the PL spectra.

Ramos et al. [61, 62] studied the photocatalytic methyl orange (MO) dye degradation performances of the ZnO and ZnO-rGO Nanorods grown from seed layers fabricated by electrospinning using different spinning voltages and amounts of rGO. The results are shown in Fig. 4.16. The photocatalyst fabricated with 40 kV shows the highest photocatalytic activity, compared with the pure ZnO and the other ZnO-rGO Nanorods obtained from seeds prepared with 20–30 kV (Fig. 4.16a). Meanwhile, the series of ZnO-rGO Nanorods fabricated with 0.2 wt.% rGO show the highest photocatalytic activity compared with the other ZnO-rGO Nanorods fabricated with 0.1 and 0.3 wt.% rGO (Fig. 4.16b). It should be noted that there is no linear dependence between the rGO amount added and the photocatalytic activity; a high content of rGO (0.3 wt.%) led to a decrement in the photocatalytic efficiency. Indeed, as previous work reported [76, 77] an excessive rGO amount, can promote the recombination of pairs in rGO, reducing the photocatalytic activity of the photocatalysts.

Furthermore, Wang et al. [78] investigated a hybrid of ZnO Nanorods grown onto three-dimensional (3D) reduced graphene oxide (RGO) modified nickel foam (ZnO/RGO@NF), by a facile hydrothermal method, for dyes degradation through photocatalysis. In particular, this sample was used as a catalyst for the degradation of 10 ppm of malachite green (MG) in seawater in a continuous flow system (Fig. 4.17). The results shown in Figs. 4.18a, b indicate that ZnO/RGO@NF samples efficiently degrade MG dye in seawater in about 15 min under UV light. In comparison, only 68% of the dye was degraded by ZnO Nanorods deposited on nickel foam (ZnO@NF), and 38% by RGO deposited on the same substrate (RGO@NF) (Fig. 4.18b). Besides, the

4.3 ZnO Nanorods-Based Carbonaceous Materials

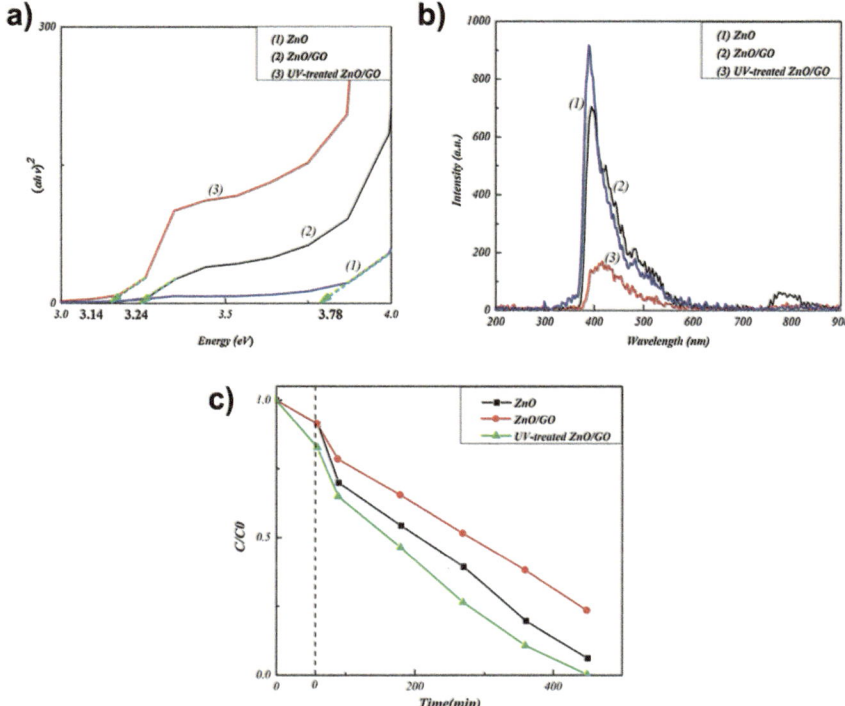

Fig. 4.15 a Tauc plots for the estimation of optical band gaps, **b** PL emission spectra, and **c** Photocatalytic degradation curves of MB for the ZnO, ZnO/GO, and UV-treated ZnO/GO Nanorods films [59]

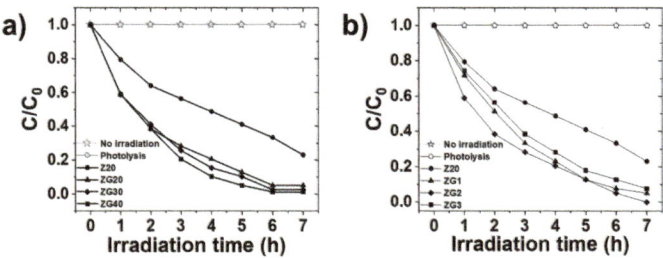

Fig. 4.16 Photodegradation curves of methyl orange over ZnO-rGO Nanorods photocatalysts grown from seed layers fabricated by electrospinning using different **a** spinning voltages and **b** amounts of rGO [61, 62]

kinetic rate of the obtained samples was analyzed according to the pseudo-first-order kinetics [79], and the degradation rate constants (k) obtained were 0.29, 0.06, and 0.01 min^{-1} for ZnO/RGO@NF, ZnO@NF, and RGO@NF, respectively (Fig. 4.18c). The rate constant of ZnO/RGO@NF was approximately 4.8 times higher than that

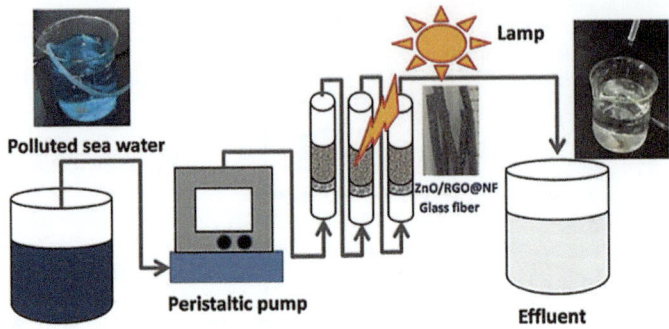

Fig. 4.17 The continuous flow system for degradation of Malachite Green in Seawater [78]

of ZnO@NF. The durability analysis of the ZnO/RGO@NF photocatalyst, shown in Fig. 4.18d, demonstrated stable photocatalytic performance, with no significant change in the degradation efficiency of MG observed during 5 h of continuous operation. This stability highlights the potential of ZnO/RGO@NF as a promising catalyst for the degradation of organic dyes in seawater.

Dai et al. [80] investigated the photocatalytic degradation of salicylic acid (SA). The reaction was performed in a quartz reactor under visible light irradiation, containing 100 mL of a 10 ppm aqueous SA solution. To this solution, 0.01 g of ZnO Nanorods/g-C_3N_4 was added. Figure 4.19a shows the degradation rate of salicylic acid by the catalysts, determined using a first-order kinetic equation. This experiment demonstrated that the best catalytic performance, indicated by the highest reaction rate constant for SA, was 0.0063 h^{-1} when 20% ZnO Nanorods/g-C_3N_4 was used as the photocatalyst, which was five times higher than the reaction rate constant of g-C_3N_4 used as the photocatalyst. This study demonstrated that the improved performance was due to the synergistic effect of the bimetallic nanoparticles. Moreover, the 20% ZnO Nanorods/g-C_3N_4 catalyst was recovered after degradation, filtered through distilled water, dried, and reused. Figure 4.19b shows the degradation rate over four recovery tests. The photodegradation efficiency of the catalyst remained consistent across these tests, indicating that the synthesized catalyst is recyclable rather than a one-time-use photocatalyst. This study will be helpful in future applications for degrading organic compounds to prevent and control environmental pollution.

The improvement in photocatalytic degradation efficiency of ZnO/graphene Nanorods compared to pristine ZnO Nanorods is attributed to the incorporation of graphene onto the ZnO nanostructures. Graphene enhances the photocatalytic activity of ZnO due to the following reasons: (i) it acts as an electron transport medium, effectively inhibiting the recombination of electrons and holes [81, 82]; (ii) it increases the adsorption of dye molecules through π-π interactions between the dye molecules and graphene [83, 84]; and (iii) it enhances light utilization efficiency [85, 86].

4.3 ZnO Nanorods-Based Carbonaceous Materials

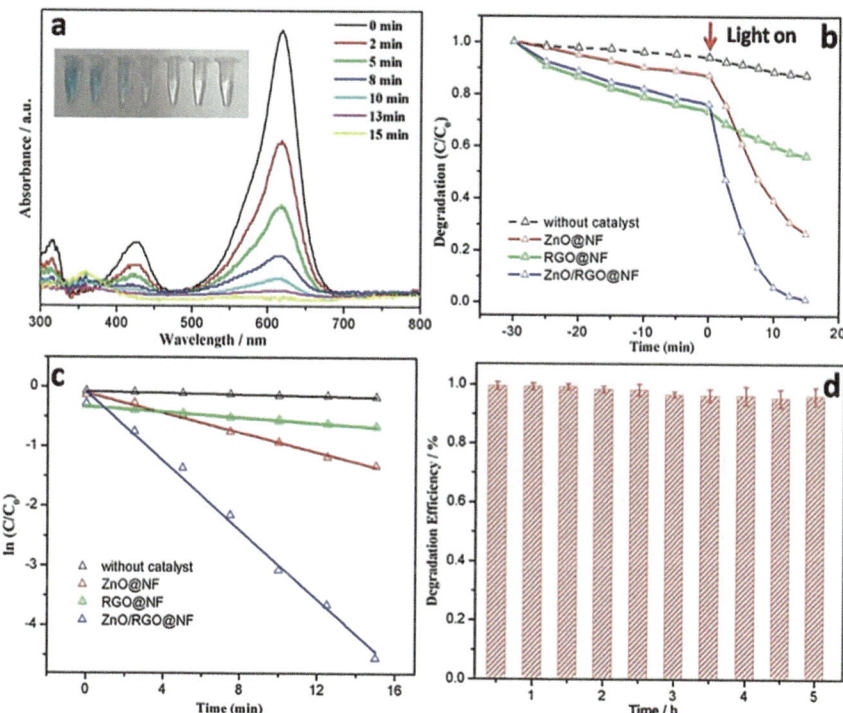

Fig. 4.18 **a** The photocatalytic degradation process of malachite green (MG) (20 mg/L) in seawater in the presence of ZnO/RGO@NF under UV irradiation (the inset photo shows the color change of MG in seawater with time); **b** MG degradation (%) at different irradiation time and **c** first order kinetics plot of $\ln(C/C_0)$ versus irradiation time for MG degradation; and, **d** Degradation efficiency of ZnO/RGO@NF for continuous 5 h degradation of MG (10 mg/L) in seawater [78]

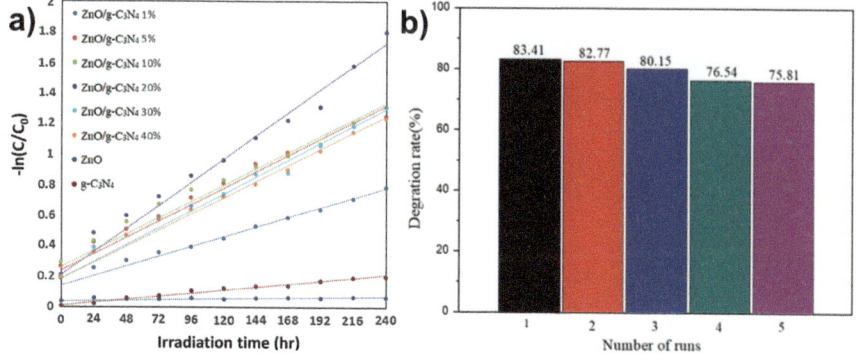

Fig. 4.19 **a** First-order kinetic plots for the photodegradation of salicylic acid (SA) and **b** the performance of SA photocatalytic degradation using ZnO Nanorods/g-C_3N_4 (20%) over multiple cycles [80]

According to all the experiments above, a possible mechanism for the photocatalytic degradation of dye molecules using ZnO-graphene Nanorods has been proposed and is shown in Fig. 4.20. Under light irradiation on the ZnO/graphene Nanorods surface, the excitation of electrons takes place from the valence band (VB) to the conduction band (CB), leaving holes in the former band to generate electron–hole pairs in ZnO NRs. Graphene sheets receive the excited electrons from the conduction band of ZnO, inhibiting the electron–hole recombination. This is mainly because graphene has a zero band gap and excellent conductivity; electrons can be transferred rapidly [87, 88]. The photogenerated electrons interact with oxygen (O_2), generating superoxide anion radicals ($\cdot O_2^-$). Meanwhile, the holes easily react with water molecules (H_2O) to form hydroxyl radicals ($\cdot OH$), and electrons produce $\cdot O_2^-$ by interacting with O_2. The generated radicals, superoxide and hydroxyl, are highly oxidative and can attack various organic dyes, degrading them into CO_2 and H_2O [89]. Moreover, the presence of graphene (GR) in the ZnO/GR Nanorods enhances the absorption of dye molecules, thereby improving the photocatalytic activity of ZnO/GR Nanorods [81, 90]. Several factors contribute to the enhanced photocatalytic performance of ZnO/GR Nanorods, including the improved synergistic interaction between the graphene layers and ZnO Nanorods, increased adsorption affinity for dye molecules, and enhanced light-harvesting capabilities [90–93].

After a detailed analysis grounded in their characterizations, Zhong et al. [94] propose a potential photocatalytic mechanism for a g-C_3N_4/ZnO Nanorods composite photocatalyst. Figure 4.21 provides a schematic view of this mechanism. Under simulated sunlight, g-C_3N_4 absorbs visible light, causing electrons in its valence band (VB) to transition to the conduction band (CB), thus creating holes in the VB. When excited electrons from the valence band (VB) of g-C_3N_4 transfer to the conduction band (CB), they directly reduce Cr(VI) to Cr(III) due to the presence of separated electrons. Because the conduction band of g-C_3N_4 is more negative

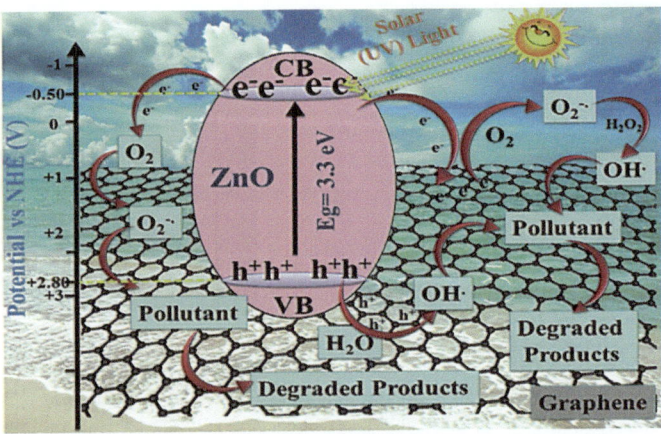

Fig. 4.20 A schematic diagram for the charge-transfer process and photocatalytic mechanism of ZnO/graphene-based Nanorods for pollutant eradication [93]

4.3 ZnO Nanorods-Based Carbonaceous Materials

than that of ZnO, the photogenerated electrons from g-C$_3$N$_4$ easily jump to the CB of ZnO. According to electron spin resonance spectroscopy experiments, the CB electrons of ZnO can reduce oxygen (O$_2$) to form highly active superoxide radical ions (•O$_2^-$), which ultimately decompose the organic dye into carbon dioxide and water. Additionally, the holes (h$^+$) in the VB of g-C$_3$N$_4$ can oxidize H$_2$O or –OH to produce hydroxyl radicals (•OH), which decompose the organic dye. Combining these findings with the results of free radical trapping experiments, it is evident that •O$_2^-$ is the most crucial active substance for degrading organic dyes, while ·OH acts as a secondary active substance. The relevant reactions are outlined in the following equations.

$$g - C_3N_4/ZnO \rightarrow g - C_3N_4/ZnO(h^+ + e^-) \tag{4.1}$$

$$ZnO(e^-) + O_2 \rightarrow \cdot O_2^- \tag{4.2}$$

$$g - C_3N_4(h^+) + H_2O/OH^- \rightarrow \cdot OH \tag{4.3}$$

$$\cdot O_2^- / \cdot OH + organic dyes \rightarrow CO_2 + H_2O \tag{4.4}$$

$$Cr_2O_7^{2-} + 14H^+ + 6e^- \rightarrow 2Cr^{3+} + 7H_2O \tag{4.5}$$

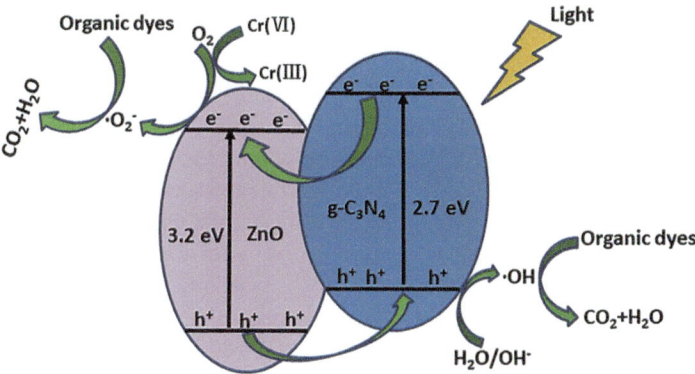

Fig. 4.21 The fundamental mechanism of photocatalysis involving g-C$_3$N$_4$/ZnO Nanorods [94]

4.4 ZnO Ternary Composites

As mentioned in previous sections, synthesizing semiconductor heterojunctions based on ZnO Nanorods has enhanced the separation of photogenerated electron–hole pairs, thereby increasing photocatalytic activity. However, there is still a challenge in these binary composites: the separated charges are often difficult to transfer and effectively participate in redox reactions. A ternary heterostructure composite system can be designed to address this limitation by introducing an additional cocatalyst or doping with a metal, non-metal, or rare earth. This cocatalyst can provide efficient sites for redox reactions, reducing the likelihood of charge recombination, and decreasing the activation energy of the response. As a result, the ternary composite offers more active sites for redox reactions and inhibits charge recombination better than binary nanocomposites, thereby improving their performance in environmental remediation and photocatalytic applications. For example, Gupta et al. [95] demonstrated that ZnO Nanorods fabricated by the hydrothermal method exhibit superior photocatalytic activity for antibacterial applications compared to combustion-synthesized nanoparticles. This enhanced performance is attributed to a higher number of surface charge carriers, dimensional anisotropy, and an efficient charge transfer together with the unidirectional structure of the Nanorods. These factors reduce recombination and enhance charge separation, boosting photocatalytic activity compared to that observed in spherical nanoparticles. Furthermore, the study introduced a novel ternary hybrid composite, ZnO Nanorods-CdS-Ag, which exhibited superior charge separation and reduced recombination compared to a physical mixture of ZnO Nanorods, CdS nanoparticles, and Ag. Incorporating CdS as a photosensitizer on the ZnO surface improved light absorption. Adding 1% Ag further enhanced the photocatalytic activity, because silver provides efficient charge transfer and thus facilitates the efficient electron–hole separation. The reactions followed first-order kinetics, with rate constants for E. coli inactivation in the presence of ZnO Nanorods-CdS-Ag measured at 11 ± 0.3 h^{-1} under UV light and 12 ± 0.6 h^{-1} under visible light. These findings confirm that ZnO Nanorods-CdS-Ag is a highly efficient material for the photocatalytic inactivation of E. coli.

On the other hand, Ramos et al. [96] investigated the use of zinc oxide/reduced graphene oxide (ZnO/rGO) nanostructures doped with transition metals (Fe, Cu, and Co) for photocatalytic applications. These nanostructures were fabricated on fluorine-doped tin oxide (FTO) substrates using an innovative, straightforward electrospinning-assisted hydrothermal method. The research investigates the effects of reduced graphene oxide (rGO) sheets attached to ZnO nanostructures and the doping of Fe, Cu, and Co ions. The photocatalytic performance of undoped and doped ZnO/rGO nanostructures was assessed by degrading Rhodamine B (RhB) dye under simulated sunlight. The improvement in photocatalytic activity compared to pure ZnO is primarily due to the introduction of dopants and the inclusion of rGO. In both undoped and doped ZnO/rGO nanostructures, the enhanced efficiency results from the strong interfacial charge transfer between ZnO and rGO, which enables the efficient separation of photogenerated carriers. Moreover, the doped ZnO/rGO samples exhibit superior performance than undoped ones because Fe^{3+}, Cu^{2+}, and Co^{2+} ions in the doped samples create additional electron–hole traps. These traps reduce recombination rates and improve photocatalytic efficiency. Among the doped samples, the Co-ZnO/rGO photocatalyst demonstrated the highest photocatalytic activity, achieving a 95% removal of RhB within 7 h of irradiation. The study highlights challenges in controlling the doping process to ensure uniform properties. Nonetheless, the potential applications and benefits of transition metal-doped ZnO/rGO ternary nanostructures make them promising for future efficient and sustainable photocatalytic applications.

Li et al. [97] prepared oxygen defect-rich pencil-like ZnO Nanorods by a simple hydrothermal method, and modified them with Ag and Carbon nanodots (CDots) co-catalysts. Characterization results showed that the pencil-like morphology and oxygen defects increased carrier concentration and enhanced charge carrier separation in ZnO. The introduction of Ag and CDots co-catalysts further improved the electron–hole pairs separation and visible light utilization of photocatalysts. Tetracycline hydrochloride (TCH) was used as a model pollutant to assess the photocatalytic performance of the synthesized catalysts. The CDots/Ag/ZnO composite demonstrated the highest TCH removal and mineralization efficiency, achieving approximately 94.95% under UV–Vis light. A possible degradation mechanism and pathways for TCH were proposed, revealing that h^+, $\cdot O_2^-$, and •OH radicals were the active species in the photocatalytic degradation process. Additionally, recycling tests confirmed the high stability of the CDots/Ag/ZnO catalyst. Finally, the biotoxicity assessment concluded that the ternary photocatalyst exhibited good biocompatibility and low biotoxicity.

A photocatalytic mechanism for the CDots/Ag/ZnO ternary photocatalyst is illustrated in Fig. 4.22. Under UV–Vis light irradiation, electrons in the ZnO rods are excited from the valence band to the conduction band. The oxygen vacancies and interstitial oxygen defects in the CDots/Ag/ZnO structure can act as electron acceptors and hole trappers, respectively. These defects further adjust the band gap and prevent the recombination of photogenerated electron–hole pairs, thereby reducing the energy loss associated with exciton recombination. Among the photocatalysts prepared, CDots/Ag/ZnO had the highest relative content of singly charged oxygen vacancies, which favored the generation of $\cdot O_2^-$ radicals [98]. Additionally, the Surface Plasmon Resonance (SPR) effect of Ag and the upconversion luminescent properties of CDots worked together to enhance the visible light utilization of the CDots/Ag/ZnO ternary photocatalyst. The hot electrons generated by the SPR effect of Ag nanoparticles, combined with electrons transferred from ZnO rods, were captured by oxidizing agents in the solution. The accumulated photogenerated electrons could then be stored and shuttled by the CDots. As a result, the interface of the CDots/Ag/ZnO photocatalyst exhibited enhanced separation of photogenerated electron–hole pairs. The photogenerated electrons in the conduction band of CDots/Ag/ZnO could be captured by O_2 to generate $\cdot O_2^-$ radicals, as the flat band potential (−0.68 eV vs. NHE) is more negative than the redox potential of $O_2/\cdot O_2^-$ (−0.046 eV vs. NHE) [99]. Furthermore, the valence band bottom potential (2.40 eV vs. NHE) is more positive than the redox potential of $OH^-/\cdot OH$ (2.38 eV vs. NHE), resulting in the effective photocatalytic degradation of TCH. The reaction can be described as follows:

$$ZnO + h\nu \rightarrow ZnO\left(e^-_{CB} + h^+_{VB}\right) \tag{4.6}$$

$$Ag + h\nu \rightarrow Ag\left(e^-_{hot}\right) \tag{4.7}$$

$$Ag\left(e^-_{hot}\right) \rightarrow ZnO\left(e^-_{CB}\right) \tag{4.8}$$

$$CDots + h\nu \rightarrow CDots\left(e^-_{upconverted}\right) \tag{4.9}$$

$$CDots + e^-_{CB}(from ZnO/Ag) \rightarrow Stored\, electrons \tag{4.10}$$

$$Stored\, electrons \rightarrow Transferred\, to\, O_2 \tag{4.11}$$

$$O_2 + e^-_{CB}(from ZnO/Ag/CDots) \rightarrow \cdot O_2^- \tag{4.12}$$

$$h^+_{VB} + OH^- \rightarrow \cdot OH \tag{4.13}$$

$$h^+_{VB} + H_2O \rightarrow \cdot OH + H^+ \tag{4.14}$$

$$TCH + \cdot O_2^- / \cdot OH \rightarrow Degraded\, Products \tag{4.15}$$

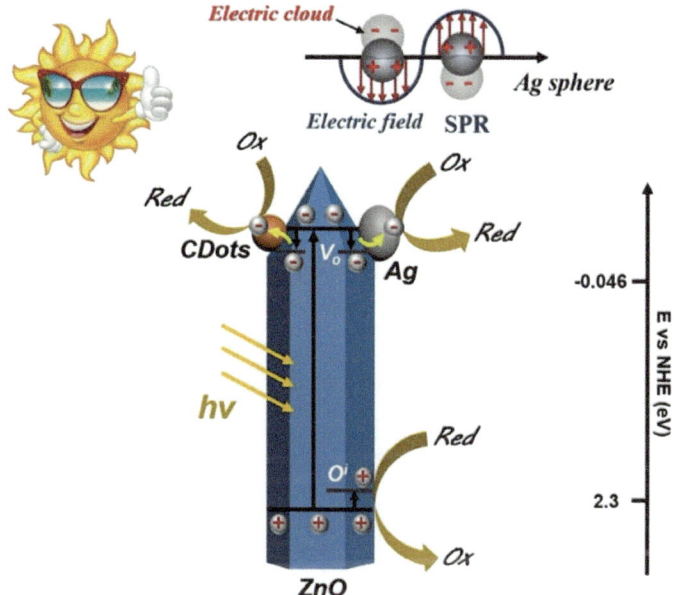

Fig. 4.22 A photocatalytic mechanism for the CDots/Ag/ZnO ternary photocatalyst [97]

References

1. A. Šutka et al., Co doped ZnO nanowires as visible light photocatalysts. Solid State Sci. **56**, 54–62 (2016). https://doi.org/10.1016/j.solidstatesciences.2016.04.008
2. B. Poornaprakash, U. Chalapathi, K. Subramanyam, S.V.P. Vattikuti, S.H. Park, Wurtzite phase Co-doped ZnO nanorods: Morphological, structural, optical, magnetic, and enhanced photocatalytic characteristics. Ceram. Int. **46**(3), 2931–2939 (2020). https://doi.org/10.1016/j.ceramint.2019.09.289
3. L. Roza, Y. Febrianti, S. Iwan, V. Fauzia, The role of cobalt doping on the photocatalytic activity enhancement of ZnO nanorods under UV light irradiation. Surf.S Interfaces **18**, 100435 (2020). https://doi.org/10.1016/j.surfin.2020.100435
4. M. Shirzad-Siboni, A. Jonidi-Jafari, M. Farzadkia, A. Esrafili, M. Gholami, Enhancement of photocatalytic activity of Cu-doped ZnO nanorods for the degradation of an insecticide: Kinetics and reaction pathways. J. Environ. Manage. **186**, 1–11 (2017). https://doi.org/10.1016/j.jenvman.2016.10.049
5. N.A. Putri, V. Fauzia, S. Iwan, L. Roza, A.A. Umar, S. Budi, Mn-doping-induced photocatalytic activity enhancement of ZnO nanorods prepared on glass substrates. Appl. Surf. Sci. **439**, 285–297 (2018). https://doi.org/10.1016/j.apsusc.2017.12.246
6. C. Han, L. Duan, X. Zhao, Z. Hu, Y. Niu, W. Geng, Effect of Fe doping on structural and optical properties of ZnO films and nanorods. J. Alloy. Compd. **770**, 854–863 (2019). https://doi.org/10.1016/j.jallcom.2018.08.217
7. N.R. Khalid et al., Enhanced photocatalytic activity of Al and Fe co-doped ZnO nanorods for methylene blue degradation. Ceram. Int. **45**(17), 21430–21435 (2019). https://doi.org/10.1016/j.ceramint.2019.07.132
8. M. Tosun, S.D. Senol, L. Arda, Effect of Mn/Cu co-doping on the structural, optical and photocatalytic properties of ZnO nanorods. J. Mol. Struct. **1212**, 128071 (2020). https://doi.org/10.1016/j.molstruc.2020.128071

9. M. Khatamian, A.A. Khandar, B. Divband, M. Haghighi, S. Ebrahimiasl, Heterogeneous photocatalytic degradation of 4-nitrophenol in aqueous suspension by Ln (La^{3+}, Nd^{3+} or Sm^{3+}) doped ZnO nanoparticles. J. Mol. Catal. A: Chem. **365**, 120–127 (2012). https://doi.org/10.1016/j.molcata.2012.08.018
10. D. Ranjith Kumar, K.S. Ranjith, and R.T. Rajendra Kumar, Structural, optical, photocurrent and solar driven photocatalytic properties of vertically aligned samarium doped ZnO nanorod arrays. Optik. **154**, 115–125 (2018). https://doi.org/10.1016/j.ijleo.2017.10.004
11. S.P. Meshram, P.V. Adhyapak, S.K. Pardeshi, I.S. Mulla, and D.P. Amalnerkar, Sonochemically generated cerium doped ZnO nanorods for highly efficient photocatalytic dye degradation. Powder Technol. **318**, 120–127 (2017). https://doi.org/10.1016/j.powtec.2017.05.044
12. P.V. Korake, A.N. Kadam, and K.M. Garadkar, Photocatalytic activity of Eu^{3+}-doped ZnO nanorods synthesized via microwave assisted technique. J. Rare Earths. **32**(4), 306–313 (2014). https://doi.org/10.1016/S1002-0721(14)60072-7
13. P.V. Korake, R.S. Dhabbe, A.N. Kadam, Y.B. Gaikwad, and K.M. Garadkar, Highly active lanthanum doped ZnO nanorods for photodegradation of metasystox. J. Photochem. Photobiol. B: Biol. **130**, 11–19 (2014). https://doi.org/10.1016/j.jphotobiol.2013.10.012
14. X. Liu, F. Fu, H. Zuo, Lanthanum ions-induced synthesis of ZnO nanostructures from zinc foil: Morphology change and photocatalytic activity. Surf.S Interfaces **1–3**, 29–34 (2016). https://doi.org/10.1016/j.surfin.2016.07.006
15. Y. Luo et al., Fabrication and photocatalytic properties of Gd-doped ZnO nanoparticle-assembled nanorods. Mater. Lett. **149**, 70–73 (2015). https://doi.org/10.1016/j.matlet.2015.02.126
16. F. Lu, J. Wang, Z. Chang, J. Zeng, Uniform deposition of Ag nanoparticles on ZnO nanorod arrays grown on polyimide/Ag nanofibers by electrospinning, hydrothermal, and photoreduction processes. Mater. Des. **181**, 108069 (2019). https://doi.org/10.1016/j.matdes.2019.108069
17. H. Zhai, X. Liu, Z. Wang, Y. Liu, Z. Zheng, X. Qin, X. Zhang, P. Wang, B. Huang, ZnO nanorod decorated by Au-Ag alloy with greatly increased activity for photocatalytic ethylene oxidation. Chin. J. Catal. **41**, 1613–1621 (2020). https://doi.org/10.1016/S1872-2067(19)63473-X
18. N.L. Gavade, A.N. Kadam, S.B. Babar, A.D. Gophane, K.M. Garadkar, S.W. Lee, Biogenic synthesis of gold-anchored ZnO nanorods as photocatalyst for sunlight-induced degradation of dye effluent and its toxicity assessment. Ceram. Int. **46**, 11317–11327 (2020). https://doi.org/10.1016/j.ceramint.2020.01.161
19. M. Sakir, S. Salem, S.T. Sanduvac, E. Sahmetlioglu, G. Sarp, M.S. Onses, E. Yilmaz, Photocatalytic green fabrication of Au nanoparticles on ZnO nanorods modified membrane as flexible and photocatalytic active reusable SERS substrates. Colloids Surf. A **585**, 124088 (2020). https://doi.org/10.1016/j.colsurfa.2019.124088
20. B. Baruah, L. Downer, D. Agyeman, Fabric-based composite materials containing ZnO-NRs and ZnO-NRs-AuNPs and their application in photocatalysis. Mater. Chem. Phys. **231**, 252–259 (2019). https://doi.org/10.1016/j.matchemphys.2019.04.006
21. S.J. Young, Y.L. Chu, Platinum nanoparticle-decorated ZnO nanorods improved the performance of methanol gas sensor. J. Electrochem. Soc. **167**, 147508 (2020). https://doi.org/10.1149/1945-7111/abc4be
22. Z. Wu, Y. Xue, H. Wang, Y. Wu, H. Yu, ZnO nanorods/Pt and ZnO nanorods/Ag heteronanostructure arrays with enhanced photocatalytic degradation of dyes. RSC Adv. **4**, 59009–59016 (2014). https://doi.org/10.1039/C4RA10753E
23. N. Alahmadi, M.S. Amin, R.M. Mohamed, Superficial visible-light-responsive Pt@ZnO nanorods photocatalysts for effective remediation of ciprofloxacin in water. J. Nanopart. Res. **22**, 1–14 (2020). https://doi.org/10.1007/s11051-020-04968-7
24. M. Arifin, L. Roza, V. Fauzia, Bayberry-like Pt nanoparticle decorated ZnO nanorods for the photocatalytic application. Results in Physics **15**, 102678 (2019). https://doi.org/10.1016/j.rinp.2019.102678
25. L. Sanchez, C. Castillo, W. Cruz, B. Yauri, M. Sosa, C. Luyo, R. Candal, S. Ponce, J.M. Rodriguez, ZnO (Ag-N) nanorods films optimized for photocatalytic water purification. Coatings **9**, 767 (2019). https://doi.org/10.3390/coatings9110767

26. M. Chakraborty, R. Thangavel, and K. Asokan, N doped ZnO and ZnO nanorods based p-n homojunction fabricated by ion implantation, AIP Conference Proceedings. **1953**, 050047 (2018). https://doi.org/10.1063/1.5032702
27. M. Wang, F. Ren, J. Zhou, G. Cai, L. Cai, Y. Hu, D. Wang, Y. Liu, L. Guo, S. Shen, N Doping to ZnO Nanorods for photoelectrochemical water splitting under visible light: engineered impurity distribution and terraced band structure. Sci. Rep. **5**, 12925 (2015). https://doi.org/10.1038/srep12925
28. A.S. Ismail, M.H. Mamat, N.D.M. Sin, M.F. Malek, S.A. Saidi, M.M. Yusoff, and M. Rusop, Structural and optical properties of N-doped ZnO nanorod arrays prepared using sol-gel immersion method. In: 2016 IEEE student conference on research and development (SCOReD), **1**, 1–6 (2016). https://doi.org/10.1109/SCORED.2016.7810067
29. N.R. Panda, B.S. Acharya, P. Nayak, Sonochemical synthesis of nitrogen doped ZnO nanorods: effect of anions on growth and optical properties. J. Mater. Sci. Mater. Electron. **24**, 4043–4049 (2013). https://doi.org/10.1007/s10854-013-1359-z
30. X. Duan, G. Chen, P. Gao, W. Jin, X. Ma, Y. Yin, L. Guo, H. Ye, Y. Zhu, J. Yu, Y. Wu, Crystallography facet tailoring of carbon doped ZnO nanorods via selective etching. Appl. Surf. Sci. **406**, 186–191 (2017). https://doi.org/10.1016/j.apsusc.2017.02.119
31. P.M. Perillo, M.N. Atia, C-doped ZnO nanorods for photocatalytic degradation of p-aminobenzoic acid under sunlight. Nano-Struct. Nano-Objects **10**, 125–130 (2017). https://doi.org/10.1016/j.nanoso.2017.04.001
32. Z. Mirzaeifard, Z. Shariatinia, M. Jourshabani, and S.M. Rezaei Darvishi, ZnO Photocatalyst revisited: effective photocatalytic degradation of emerging contaminants using S-doped ZnO nanoparticles under visible light radiation. Ind. Eng. Chem. Res. **59**, 15894–15911 (2020). https://doi.org/10.1021/acs.iecr.0c03192
33. A. Khan, M.I. Ahmed, A. Adam, A.M. Azad, M. Qamar, A novel fabrication methodology for sulfur-doped ZnO nanorods as an active photoanode for improved water oxidation in visible-light regime. Nanotechnology **28**, 055602 (2016). https://doi.org/10.1088/1361-6528/aa51b6
34. M. Hsu, C.J. Chang, S-doped ZnO nanorods on stainless-steel wire mesh as immobilized hierarchical photocatalysts for photocatalytic H_2 production. Int. J. Hydrogen Energy **39**, 16524–16533 (2014). https://doi.org/10.1016/j.ijhydene.2014.02.110
35. X. Zhang, X. Yan, J. Zhao, Z. Qin, Y. Zhang, Structure and photoluminescence of S-doped ZnO nanorod arrays. Mater. Lett. **63**, 444–446 (2009). https://doi.org/10.1016/j.matlet.2008.11.006
36. R. Saffari, Z. Shariatinia, M. Jourshabani, Synthesis and photocatalytic degradation activities of phosphorus containing ZnO microparticles under visible light irradiation for water treatment applications. Environ. Pollut. **259**, 113902 (2020). https://doi.org/10.1016/j.envpol.2019.113902
37. S. Swathi, R. Yuvakkumar, G. Ravi, E.S. Babu, D. Velauthapillai, and S.A. Alharbi, Morphological exploration of chemical vapor–deposited P-doped ZnO nanorods for efficient photoelectrochemical water splitting. **47**, 6521–6527 (2021). https://doi.org/10.1016/j.ceramint.2020.10.237
38. C. Cao, B. Zhang, and S. Lin, p-type ZnO for photocatalytic water splitting. **10**, 030901, (2022). https://doi.org/10.1063/5.0083753
39. R. Yang, F. Wang, J. Lu, Y. Lu, B. Lu, S. Li, and Z. Ye, ZnO with p-type doping: recent approaches and applications. **5**, 4014–4034 (2023). https://doi.org/10.1021/acsaelm.3c00515
40. Y. Lv, J. Liu, Z. Zhang, W. Zhang, A. Wang, F. Tian, W. Zhao, J. Yan, Green synthesis of CuO nanoparticles-loaded ZnO nanowires arrays with enhanced photocatalytic activity. Mater. Chem. Phys. **267**, 124703 (2021). https://doi.org/10.1016/j.matchemphys.2021.124703
41. X. Wang, G. Liu, G.Q. Lu, H.-M. Cheng, Stable photocatalytic hydrogen evolution from water over ZnO-CdS core-shell nanorods. Int. J. Hydrogen Energy **35**, 8199–8205 (2010). https://doi.org/10.1016/j.ijhydene.2009.12.091
42. X. Liu, J. Cao, B. Feng, L. Yang, M. Wei, H. Zhai, H. Liu, Y. Sui, J. Yang, Y. Liu, Facile fabrication and photocatalytic properties of ZnO nanorods/ZnSe nanosheets heterostructure. Superlattices Microstruct. **83**, 447–458 (2015). https://doi.org/10.1016/j.spmi.2015.03.051

43. M. Gholamveysi et al., MoSe$_2$ nanoflakes decorated ZnO nanorods: an effective photo-electrode with S-scheme heterojunction for photoelectrocatalytic degradation of tetracycline and rhodamine B. Surf.S Interfaces **40**, 103146 (2023). https://doi.org/10.1016/j.surfin.2023.103146
44. K.S. Sajiv, G. Gopakumar, M. Shanmugam, Integrated photo-absorption and improved charge transport kinetics in atomically thin MoSe$_2$-incorporated nanostructured ZnO photo-anodes for dye-sensitized solar cells. Appl. Phys. A **127**, 966 (2021). https://doi.org/10.1007/s00339-021-05127-y
45. P.G. Ramos et al., Enhanced photoelectrochemical performance and photocatalytic activity of ZnO/TiO$_2$ nanostructures fabricated by an electrostatically modified electrospinning. Appl. Surf. Sci. **426**, 844–851 (2017). https://doi.org/10.1016/j.apsusc.2017.07.218
46. A.R.B.M. Yusoff, L. Dai, H.M. Cheng, J. Liu, Graphene-based energy devices. Nanoscale **7**, 6881–6882 (2015). https://doi.org/10.1039/C5NR90062J
47. J.L. Lopes, M.J. Martins, H.I.S. Nogueira, A.C. Estrada, T. Trindade, Carbon-based heterogeneous photocatalysts for water cleaning technologies: a review. Environ. Chem. Lett. **19**, 643–668 (2021). https://doi.org/10.1007/s10311-020-01092-9
48. A. Raza, S. Altaf, S. Ali, M. Ikram, G. Li, Recent advances in carbonaceous sustainable nanomaterials for wastewater treatments. Sustain. Mater. Technol. **32**, e00406 (2022). https://doi.org/10.1016/j.susmat.2022.e00406
49. J. Phiri, P. Gane, T.C. Maloney, General overview of graphene: Production, properties, and application in polymer composites. Mater. Sci. Eng. B **215**, 9–28 (2017). https://doi.org/10.1016/j.mseb.2016.10.004
50. K. Anand, O. Singh, R.C. Singh, Different strategies for the synthesis of graphene/ZnO composite and its photocatalytic properties. Appl. Phys. A **116**, 1141–1148 (2013). https://doi.org/10.1007/s00339-013-8198-x
51. N.P.F. Gonçalves, M.A.O. Lourenço, S.R. Baleuri, S. Bianco, P. Jagdale, P. Calza, Biochar waste-based ZnO materials as highly efficient photocatalysts for water treatment. J. Environ. Chem. Eng. **10**, 107256 (2022). https://doi.org/10.1016/j.jece.2022.107256
52. I. Boukhoubza, M. Khenfouch, M. Achehboune, B.M. Mothudi, I. Zorkani, A. Jorio, Graphene oxide/ZnO nanorods/graphene oxide sandwich structure: The origins and mechanisms of photoluminescence. J. Alloy. Compd. **797**, 1320–1326 (2019). https://doi.org/10.1016/j.jallcom.2019.04.266
53. B.A. Vessalli, C.A. Zito, T.M. Perfecto, D.P. Volanti, T. Mazon, ZnO nanorods/graphene oxide sheets prepared by chemical bath deposition for volatile organic compounds detection. J. Alloy. Compd. **696**, 996–1003 (2017). https://doi.org/10.1016/j.jallcom.2016.12.075
54. M. Honda, R. Okumura, Y. Ichikawa, Direct growth of densely aligned ZnO nanorods on graphene. Jpn. J. Appl. Phys. **55**, 080301 (2016). https://doi.org/10.7567/JJAP.55.080301
55. V.Q. Dang, T.Q. Trung, D.I. Kim, L.T. Duy, B.-U. Hwang, D.-W. Lee, B.-Y. Kim, L.D. Toan, N.-E. Lee, Ultrahigh responsivity in graphene-ZnO nanorod hybrid UV photodetector. Small **11**, 3054–3065 (2015). https://doi.org/10.1002/smll.201403625
56. I. Boukhoubza, M. Khenfouch, M. Achehboune, B. Mouthudi, I. Zorkani, A. Jorio, Synthesis and characterization of Graphene oxide/Zinc oxide nanorods sandwich structure. J. Phys. Conf. Ser. **984**, 012005 (2018). https://doi.org/10.1088/1742-6596/984/1/012005
57. T.-Y. Yu, M.R. Wei, C.Y. Weng, W.M. Su, C.C. Lu, Y.T. Chen, H. Chen, Modulating the size of ZnO nanorods on SiO$_2$ substrates by incorporating reduced graphene oxide into the seed layer solution. AIP Adv. **7**, 065110 (2017). https://doi.org/10.1063/1.4986759
58. A.A.A. Mohammed, A.B. Suriani, A.R. Jabur, The enhancement of UV sensor response by zinc oxide nanorods/reduced graphene oxide bilayer nanocomposites film. J. Phys. Conf. Ser. **1003**, 012070 (2018). https://doi.org/10.1088/1742-6596/1003/1/012070
59. E. Rokhsat and O. Akhavan, Improving the photocatalytic activity of graphene oxide/ZnO nanorod films by UV irradiation, App Surface Sci. **371**, 590–595 (2016). https://doi.org/10.1016/j.apsusc.2016.02.222
60. R. Zou, G. He, K. Xu, Q. Liu, Z. Zhang, and J. Hu, ZnO nanorods on reduced graphene sheets with excellent field emission, gas sensor, and photocatalytic properties, J. Mat. Chem. A. **1**, 8445–8452 (2013). https://doi.org/10.1039/C3TA11490B

References

61. P.G. Ramos, E. Flores, C. Luyo, L.A. Sánchez, J. Rodriguez, Fabrication of ZnO-RGO nanorods by electrospinning assisted hydrothermal method with enhanced photocatalytic activity. Mater Today Commun. **19**, 407–412 (2019). https://doi.org/10.1016/j.mtcomm.2019.03.010
62. P.G. Ramos, C. Luyo, L.A. Sánchez, E.D. Gomez, J.M. Rodriguez, The spinning voltage influence on the growth of ZnO-rGO nanorods for photocatalytic degradation of methyl orange dye. Catalysts **10**, 660 (2020). https://doi.org/10.3390/catal10060660
63. J. Rodríguez, G. Feuillet, F. Donatini, D. Onna, L. Sanchez, R. Candal, M.C. Marchi, S.A. Bilmes, F. Chandezon, Influence of the spray pyrolysis seeding and growth parameters on the structure and optical properties of ZnO nanorod arrays. Mater. Chem. Phys. **151**, 378–384 (2015). https://doi.org/10.1016/j.matchemphys.2014.12.013
64. T. Majumder, S.P. Mondal, Graphene quantum dots as a green photosensitizer with carbon-doped ZnO nanorods for quantum-dot-sensitized solar cell applications. Bull. Mater. Sci. **42**, 1–5 (2019). https://doi.org/10.1007/s12034-019-1755-y
65. K. Rahimi, A. Yazdani, M. Ahmadirad, Facile preparation of zinc oxide nanorods surrounded by graphene quantum dots both synthesized via separate pyrolysis procedures for photocatalyst application. Mater. Res. Bull. **98**, 148–154 (2018). https://doi.org/10.1016/j.materresbull.2017.10.014
66. E. Haque, J. Kim, V. Malgras, K.R. Reddy, A.C. Ward, J. You, Y. Bando, Md.S.A. Hossain, Y. Yamauchi, Recent advances in graphene quantum dots: synthesis, properties, and applications. Small Methods **2**, 1800050 (2018). https://doi.org/10.1002/smtd.201800050
67. X.T. Zheng, A. Ananthanarayanan, K.Q. Luo, P. Chen, Glowing graphene quantum dots and carbon dots: properties. Synth. Biol. Appl. Small **11**, 1620–1636 (2014). https://doi.org/10.1002/smll.201402648
68. K. Rahimi, A. Yazdani, Ethanol-sensitive nearly aligned ZnO nanorod thin films covered by graphene quantum dots. Mater. Lett. **228**, 65–67 (2018). https://doi.org/10.1016/j.matlet.2018.05.137
69. K. Rahimi, A. Yazdani, M. Ahmadirad, Graphene quantum dots enhance UV photoresponsivity and surface-related sensing speed of zinc oxide nanorod thin films. Mater. Des. **140**, 222–230 (2018). https://doi.org/10.1016/j.matdes.2017.12.010
70. A. Kathalingam, H.M.S. Ajmal, D. Vikraman, S.D. Kim, H.C. Park, H.-S. Kim, Graphene quantum dots-wrapped vertically aligned zinc oxide nanorods arrays for photosensing application. J. Alloy. Compd. **853**, 157025 (2021). https://doi.org/10.1016/j.jallcom.2020.157025
71. B. Albiss, M. Abu-Dalo, Photocatalytic Degradation of Methylene Blue Using Zinc Oxide Nanorods Grown on Activated Carbon Fibers. Sustainability **13**, 4729 (2021). https://doi.org/10.1016/j.jallcom.2020.157025
72. H. Li et al., Enhanced photocatalytic activity and synthesis of ZnO nanorods/MoS$_2$ composites. Superlattices Microstruct. **117**, 336–341 (2018). https://doi.org/10.1016/j.spmi.2018.03.028
73. K. Rahimi, A. Yazdani, Incremental photocatalytic reduction of graphene oxide on vertical ZnO nanorods for ultraviolet sensing. Mater. Lett. **262**, 127078 (2020). https://doi.org/10.1016/j.matlet.2019.127078
74. G. Anoop, V. Panwar, T.Y. Kim, and J.Y. Jo, Resistive switching in ZnO nanorods/graphene oxide hybrid multilayer structures, Adv. Electron. Mater. **3**, 1600418 (2017). https://doi.org/10.1002/aelm.201600418
75. W. Kang, X. Jimeng, W. Xitao, The effects of ZnO morphology on photocatalytic efficiency of ZnO/RGO nanocomposites. Appl. Surf. Sci. **360**, 270–275 (2016). https://doi.org/10.1016/j.apsusc.2015.10.190
76. D. Fang, X. Li, H. Liu, W. Xu, M. Jiang, W. Li, X. Fan, BiVO$_4$-rGO with a novel structure on steel fabric used as high-performance photocatalysts. Sci. Rep. **7**, 7979 (2017). https://doi.org/10.1038/s41598-017-07342-1
77. H. Wu, S. Lin, C. Chen, W. Liang, X. Liu, H. Yang, A new ZnO/rGO/polyaniline ternary nanocomposite as photocatalyst with improved photocatalytic activity. Mater. Res. Bull. **83**, 434–441 (2016). https://doi.org/10.1016/j.materresbull.2016.06.036

78. Q. Wang et al., Efficient photocatalytic degradation of malachite green in seawater by the hybrid of zinc-oxide nanorods grown on three-dimensional (3D) reduced graphene oxide (RGO)/Ni foam. Materials **11**, 1004 (2018). https://doi.org/10.3390/ma11061004
79. E.H. Umukoro, M.G. Peleyeju, J.C. Ngila, and O.A. Arotiba, photoelectrochemical degradation of orange II dye in wastewater at a silver-zinc oxide/reduced graphene oxide nanocomposite photoanode, RSC Adv. **6**, 52868–52877 (2016). https://doi.org/10.1039/C6RA04156F
80. E.H. Umukoro, M.G. Peleyeju, J.C. Ngila, O.A. Arotiba, Construction of heterostructured nanorods-like ZnO/g-C$_3$N$_4$ nanocomposite to obtain enhanced photocatalytic properties. Opt. Mater. **148**, 114925 (2014). https://doi.org/10.1016/j.optmat.2024.114925
81. H. Moussa, E. Girot, K. Mozet, H. Alem, G. Medjahdi, and R. Schneider, ZnO rods/reduced graphene oxide composites prepared via a solvothermal reaction for efficient sunlight-driven photocatalysis, Appl. Catal. B: Environ. **185**, 11–21 (2016). https://doi.org/10.1016/j.apcatb.2015.12.007
82. F. Khurshid, M. Jeyavelan, M.S.L. Hudson, S. Nagarajan, Ag-doped ZnO nanorods embedded reduced graphene oxide nanocomposite for photo-electrochemical applications. R. Soc. Open Sci. **6**, 181764 (2019). https://doi.org/10.1098/rsos.181764
83. C.R. Minitha, M. Lalitha, Y.L. Jeyachandran, L. Senthilkumar, and R.T. Rajendra Kumar, Adsorption behaviour of reduced graphene oxide towards cationic and anionic dyes: Co-action of electrostatic and π–π interactions, Mater. Chem. Phys. **194**, 243–252 (2017). https://doi.org/10.1016/j.matchemphys.2017.03.048
84. B. Mao, B. Sidhureddy, A.R. Thiruppathi, P.C. Wood, A. Chen, Efficient dye removal and separation based on graphene oxide nanomaterials. New J. Chem. **44**, 4519–4528 (2020). https://doi.org/10.1039/C9NJ05895H
85. X. Men, H. Chen, K. Chang, X. Fang, C. Wu, W. Qin, S. Yin, Three-dimensional free-standing ZnO/graphene composite foam for photocurrent generation and photocatalytic activity. Appl. Catal. B **187**, 367–374 (2016). https://doi.org/10.1016/j.apcatb.2016.01.052
86. M. Xie, D. Zhang, Y. Wang, Y. Zhao, Facile fabrication of ZnO nanorods modified with RGO for enhanced photodecomposition of dyes. Colloids Surf. A **603**, 125247 (2020). https://doi.org/10.1016/j.colsurfa.2020.125247
87. K.M. Yam, N. Guo, Z. Jiang, S. Li, C. Zhang, Graphene-based heterogeneous catalysis: role of graphene. Catalysts **10**, 53 (2020). https://doi.org/10.3390/catal10010053
88. A.F. Ghanem, A.A. Badawy, M.E. Mohram, and M.H. Abdel Rehim, Synergistic effect of zinc oxide nanorods on the photocatalytic performance and the biological activity of graphene nano sheets, Heliyon. **6**, e03283 (2020). https://doi.org/10.1016/j.heliyon.2020.e03283
89. J. Wang, S. Wang, Reactive species in advanced oxidation processes: formation, identification and reaction mechanism. Chem. Eng. J. **401**, 126158 (2020). https://doi.org/10.1016/j.cej.2020.126158
90. K. Qi, B. Cheng, J. Yu, W. Ho, Review on the improvement of the photocatalytic and antibacterial activities of ZnO. J. Alloy. Compd. **727**, 792–820 (2017). https://doi.org/10.1016/j.jallcom.2017.08.142
91. S. Abdolhosseinzadeh, H. Asgharzadeh, S. Sadighikia, and A. Khataee, UV-assisted synthesis of reduced graphene oxide–ZnO nanorod composites immobilized on Zn foil with enhanced photocatalytic performance, Res. Chem. Intermed. **42**, 4479–4496 (2016). https://doi.org/10.1007/s11164-015-2291-z
92. N. Karimizadeh, M. Babamoradi, R. Azimirad, M. Khajeh, Synthesis of three-dimensional multilayer graphene Foam/ZnO nanorod composites and their photocatalyst application. J. Electron. Mater. **47**, 5452–5457 (2018). https://doi.org/10.1007/s11664-018-6427-y
93. P. Raizada, A. Sudhaik, P. Singh, Photocatalytic water decontamination using graphene and ZnO coupled photocatalysts: a review. Mater. Sci. Energy Technol. **2**, 509–525 (2019). https://doi.org/10.1016/j.mset.2019.04.007
94. Q. Zhong, H. Lan, M. Zhang, H. Zhu, M. Bu, Preparation of heterostructure g-C$_3$N$_4$/ZnO nanorods for high photocatalytic activity on different pollutants (MB, RhB, Cr(VI) and eosin). Ceram. Int. **46**, 12192–12199 (2020). https://doi.org/10.1016/j.ceramint.2020.01.265

References

95. R. Gupta, N.K. Eswar, J.M. Modak, and G. Madras, Effect of morphology of zinc oxide in ZnO-CdS-Ag ternary nanocomposite towards photocatalytic inactivation of E. coli under UV and visible light, Chem. Eng. J. **307**, 966–980 (2017). https://doi.org/10.1016/j.cej.2016.08.142
96. P.G. Ramos, J. Espinoza, L.A. Sánchez, J. Rodriguez, Enhanced photocatalytic degradation of Rhodamine B employing transition metal (Fe, Cu, Co) doped ZnO/rGO nanostructures synthesized by electrospinning-hydrothermal technique. J. Alloy. Compd. **966**, 171559 (2023). https://doi.org/10.1016/j.jallcom.2023.171559
97. T. Li, Y. Liu, M. Li, J. Jang, J. Gao, S. Dong, Fabrication of oxygen defect-rich pencil-like ZnO nanorods with CDots and Ag co-enhanced photocatalytic activity for tetracycline hydrochloride degradation. Sep. Purif. Technol. **266**, 118605 (2021). https://doi.org/10.1016/j.seppur.2021.118605
98. Y. Liu, R. Wang, Z. Yang, H. Du, Y. Jiang, C. Shen, K. Liang, and A. Xu, Enhanced visible-light photocatalytic activity of Z-scheme graphitic carbon nitride/oxygen vacancy-rich zinc oxide hybrid photocatalysts, Chin. J. Catal. **36**, 2135–2144 (2015). https://doi.org/10.1016/S1872-2067(15)60985-8
99. A. Zhang, W. Wang, D. Pei, and H. Yu, Degradation of refractory pollutants under solar light irradiation by a robust and self-protected ZnO/CdS/TiO$_2$ hybrid photocatalyst. **92**, 78–86 (2016). https://doi.org/10.1016/j.watres.2016.01.045

Chapter 5
Summary and Outlook for Future Directions

This chapter explores various methods for synthesizing zinc oxide (ZnO) and modified ZnO Nanorods, highlighting their photocatalytic potential. Specifically, we examine techniques employed to enhance the photocatalytic efficiency of ZnO under light irradiation, including metal and nonmetal doping, semiconductor coupling, the incorporation of graphene-based materials, and the formation of ternary composites. Doping ZnO with nonmetal elements significantly improves photocatalytic activity. This enhancement arises from introducing electron donors or electron mediators, which can irreversibly react with positive holes, thereby inhibiting the recombination of charge carriers and preventing backward reactions. On the other hand, coupling ZnO with small optical band gap semiconductors further boosts photocatalytic efficiency through improved charge separation. Electrons injected from the small optical band gap semiconductor into the conduction band of ZnO facilitate this process, enhancing the overall performance.

Additionally, the development of ZnO ternary composites has garnered significant attention due to their enhanced photocatalytic properties. Ternary composites typically consist of ZnO combined with two additional components, including other metal oxides, carbon-based materials, or semiconductors. These composites can synergistically improve charge separation, increase light absorption, and enhance the stability of the photocatalyst. For instance, coupling ZnO with transition metal oxides, such as TiO_2 or Fe_2O_3, can result in a broader spectrum of light absorption and increased surface activity. Similarly, incorporating carbon-based materials like reduced graphene oxide (rGO) can facilitate efficient electron transfer and increase the surface area for adsorption, leading to superior photocatalytic performance.

Furthermore, integrating graphene-based nanomaterials with ZnO Nanorods has demonstrated a marked improvement in photodegradation capabilities for various pollutant dyes. This enhancement results from an expanded light absorption range and improved charge separation and adsorption capacities.

In summary, modified ZnO Nanorods have emerged as promising photocatalysts, particularly for the degradation of organic dyes, owing to their high photocatalytic activity under UV, visible, and solar irradiation, as well as their nontoxicity, photostability, straightforward synthesis procedures, and cost-effectiveness. However, a deeper understanding of the advanced photocatalytic oxidation degradation mechanisms is essential for designing efficient photoreactors and systems focused on dye degradation in wastewater streams. This advancement represents a crucial step in waste pollution management, fostering a healthier living environment for future generations.

To fully harness the commercial potential of ZnO nanorods, further research is needed to enhance large-scale production, assess economic feasibility, and ensure long-term durability. If these objectives are achieved, ZnO nanorods could significantly contribute to water treatment systems across various industries, promoting environmental sustainability and reducing pollution. Additionally, exploring the synergies between ZnO and emerging nanomaterials can lead to innovative photocatalytic solutions for pressing environmental challenges. Despite the promising capabilities of ZnO nanorods, there is limited information regarding their use in pilot plants for contaminant treatment. This knowledge gap highlights the need for extensive research and real-world trials to assess their effectiveness and scalability in wastewater treatment applications. Addressing this issue is essential for successfully implementing ZnO-based photocatalytic systems in large-scale environmental remediation projects.

GPSR Compliance

The European Union's (EU) General Product Safety Regulation (GPSR) is a set of rules that requires consumer products to be safe and our obligations to ensure this.

If you have any concerns about our products, you can contact us on ProductSafety@springernature.com

In case Publisher is established outside the EU, the EU authorized representative is:

Springer Nature Customer Service Center GmbH
Europaplatz 3
69115 Heidelberg, Germany

Batch number: 09745627

Printed by Printforce, the Netherlands